1 MONTH OF
FREE
READING

at

www.ForgottenBooks.com

By purchasing this book you are eligible for one month membership to ForgottenBooks.com, giving you unlimited access to our entire collection of over 1,000,000 titles via our web site and mobile apps.

To claim your free month visit:
www.forgottenbooks.com/free299320

ISBN 978-0-484-61173-2
PIBN 10299320

CONTRIBUTIONS

TO THE

NATURAL HISTORY

OF THE

LEPIDOPTERA

OF

NORTH AMERICA

VOL. II
No. I

ILLUSTRATIONS OF
RARE AND TYPICAL LEPIDOPTERA
(*Continued*)

BY

WILLIAM BARNES, S. B., M. D.

AND

J. H. McDUNNOUGH, Ph. D.

DECATUR, ILL.
THE REVIEW PRESS
MARCH 10, 1913

Published
Under the Patronage
of
Miss Jessie D. Gillett
Elkhart, Ill.

INTRODUCTION

The following twenty-one plates represent the balance and conclusion of those issued in Part IV of our first Volume. Owing to the inexperience of the photographer the result has been by no means satisfactory and deficiencies in the photographs have naturally been exaggerated in the half-tone reproduction. We think, however, that in future we can promise a marked improvement in the quality of our illustrations. We have of late had some excellent work done by an experienced commercial photographer and hope to publish samples of his work shortly; in the meantime we offer the following plates for what they are worth. Many of the specimens figured are readily recognizable, and for those that are not so good we crave the indulgence of the entomological public.

As in our previous publication a number of the illustrations represent types, co-types or specimens that agree with the type. A few species are undescribed; to these we have affixed the manuscript names and expect to publish a full description in a later number.

4

PLATE I

Fig. 1. MELITAEA ALMA Stkr. Pyramid Lake, Nevada, ♂.
2. MELITAEA ALMA Stkr. Pyramid Lake, Nevada. ♂ underside.
3. MELITAEA ALMA Stkr. Pyramid Lake, Nevada. ♀.
4. MELITAEA ALMA Stkr. Pyramid Lake, Nevada. ♀ underside.

The specimen figured in figure 1 has been compared with Strecker's type, which is also a ♂, and is almost an exact counterpart of this. The species has been generally confused with a suffused form of *fulvia* Edw. with which it is very similar, and until we received our Nevada material we had fallen into the same error. The specimen figured by Clemence (Can. Ent. XLIV 102, 1912) and misidentified by ourselves, is not *alma* ♀ but this suffused form of *fulvia*. The best point of distinction between the two species is found on the underside of the secondaries; in *alma* a subbasal black band is more or less indicated, being especially strong in the ♀'s; in *fulvia* this band is entirely lacking. No mention is made in Strecker's description of this band but the type ♂ distinctly shows the black costal spot and traces of black in the cell. Besides the Nevada locality, we also have received the species from Provo, Utah; it is evidently the most northern form of the *fulvia* group and approaches *leanira* in possessing the subbasal band. On the upper side the black markings are less developed than in the most suffused form of *fulvia* we have seen, and the pale yellow maculation tends to deepen in color and become lost in the ground color; this is especially noticeable in the median band and the spots towards base of wing.

5. MELITAEA FULVIA Edw. Santa Catalina Mts., Ariz. ♂ underside.
6. MELITAEA FULVIA Edw. Santa Catalina Mts., Ariz. ♀ underside.
7. MEGATHYMUS NEUMOEGENI STEPHENSI Skin. La Puerta, S. Calif. ♀.

We would call attention to the much narrower macular band of the primaries as compared with the ♀'s of the type form. The color of the spots is pale whitish yellow.

PLATE II

Fig. 1. OXYCNEMIS ERRATICA B. & McD. Brownsville, Tex. ♂ Type.

2. CERURA RARATA Wlk. Brownsville, Tex. ♂.

This species is new to our fauna. The specimen figured agrees well with Druce's figure (Biol. Cent. Am. III, Pl. 91, Fig. 7) except that it is considerably smaller.

3. PARORA SNOWI Sm. Brownsville, Tex. ♂.

One form of this variable species; other specimens before us are mottled with olivaceous and yellow.

4. TRACHEA MISELLUS Sm. Brownsville, Tex. ♂.

Apparently a race of *miseloides* Gn. as stated by Smith.

5. ACRONYCTA CONNECTA Grt. Brownsville, Tex. ♀.

We imagine our specimen to be this species, although it is less suffused with blackish than most of those in our series. If our identification is correct, this is the most southern locality from which the species has been reported.

6. ACRONYCTA LEPETITA Sm. Brownsville, Tex. ♂.

Agrees well with the description and is from the type locality.

7. TARACHE CRETATA G. & R. Brownsville, Tex. ♀.

8. LITODONTA HYDROMELI Harv. Brownsville, Tex. ♂.

A pale form with primaries largely pale green and with the subterminal purplish shading obsolescent.

9. LEUCOCNEMIS NIVALIS Sm. Brownsville, Tex. ♂.

A very strongly marked specimen with the median oblique line distinct; in other specimens the line is very faint and a few have the primaries uniform white with no traces of the line at all. In the specimen figured the primaries are largely suffused with gray scales.

10. SCHINIA BUTA Sm. Witch Creek, Calif. ♂.

11. PALEACRITA LONGICILIATA Hulst. San Diego, Calif. ♂.

12. ATHETIS MINUSCULA B. & McD. Brownsville, Tex. ♀ Co-type.

13. POLIA (MAMESTRA) ASCULA Sm. Eureka, Utah. ♂.

Agrees with ♂ co-type in Coll. Barnes.

14. POLIA ASCULA Sm. Eureka, Utah. ♂.

A more strongly marked specimen than the preceding.

15. POLIA ASCULA Sm. Eureka, Utah. ♀.

Agrees with ♀ co-type in our possession; the species is variable in regard to the distinctness of the maculation.

16. SCOTOGRAMMA (MAMESTRA) ORIDA Sm. Eureka, Utah. ♂.

17. SCOTOGRAMMA (MAMESTRA) ORIDA Sm. Eureka, Utah. ♀.

Agrees with 2 ♀ co-types in Coll. Barnes.

18. OZODANIA SUBRUFA B. & McD. Brownsville, Texas. ♀ Co-type.

PLATE II

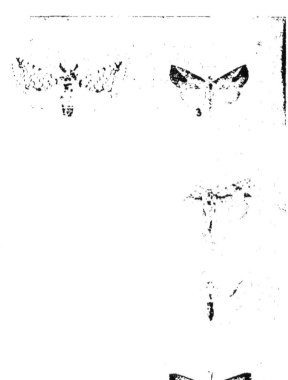

13

16

8

PLATE III

Fig. 1. EUMICHTIS (HADENA) LODA Stkr. Duncans, Vanc. Is. ♂.
2. EUMICHTIS (HADENA) LODA Stkr. Duncans, Vanc. Is. ♀.
Agrees with the type in the Strecker Coll.
3. GORTYNA (HELOTROPHA) RENIFORMIS Grt. Duncans, Vanc. Is. ♀.
4. POLIA (MAMESTRA) CIRCUMCINCTA Sm. Plumas Co., Calif. ♂.
Agrees with a specimen that has been compared with the type.
5. EUXOA STRIGILIS Grt. Duncans, Vanc. Is. ♀.
This specimen is not typical, being much more contrasting in coloration. The species is, however, very variable.
6. FISHIA HANHAMI Sm. Duncans, Vanc. Is. ♂.
Agrees with ♂ co-type in Coll. Barnes.
7. FISHIA HANHAMI Sm. Duncans, Vanc. Is. ♀.
8. POLIA (MAMESTRA) TUANA Sm. Palmerlee, Ariz. ♀.
Agrees with co-types in Coll. Barnes.
9. EUXOA DECOLOR Morr. Duncans, Vanc. Is. ♀.
A very bright brown contrasting specimen.
10. POLIA TUANA Sm. Palmerlee, Ariz. ♂.
11. POLIA NOVERCA Grt. Glenwood Spgs., Colo. ♂.
Closely allied to *tuana* Sm. but lacks claviform and the white outer edging to reniform.
12. CONISTRA (SCOPELOSOMA) TRISTIGMATA Grt. Duncans, Vanc. Is. ♂.
13. POLIA NOVERCA Grt. Glenwood Spgs., Colo. ♀.
14. METANEMA HIRSUTARIA B. & McD. San Diego, Calif. ♂ Co-type.
A rather aberrant specimen.
15. PERIGRAPHA (STRETCHIA) TRANSPARENS Grt. Duncans, Vanc. Is. ♀.
A very dark deep purple-red suffused form.
16. METANEMA HIRSUTARIA B. & McD. San Diego, Calif. ♂ Type.

PLATE III

PLATE IV

Fig. 1. SCHINIA JAGUARINA Gn. Denver, Colo. ♂.
2. SCHINIA JAGUARINA Gn. White Mts., Ariz. ♀.
3. SCHINIA JAGUARINA Gn. Southern Pines, N. C. ♀.
This is apparently a wide-spread and rather variable species; our South. Pines specimens are all much brighter in coloration and generally more contrasting; specimens from Florida seem to intergrade between the two forms.
4. SCHINIA BIFORMA Sm. Kerrville, Tex. ♀ Co-type.
5. SCHINIA PETULANS Hy. Edw. Ft. Meade, Fla. ♀.
Agrees well with the description and Hampson's figure; we have not seen the type.
6. SCHINIA OCULATA Sm. Redington, Ariz. ♂.
Agrees with the ♂ type in Coll. Barnes.
7. SCHINIA RENIFORMIS Sm. Deming, N. M. ♀.
Agrees with the ♀ co-type in Coll. Barnes.
8. POLIA (MAMESTRA) ORTRUDA Sm. Glenwood Sp., Colo. ♂ Paratype.
9. PODAGRA CRASSIPES Sm. Yuma Co., Ariz. ♀ Co-type.
10. POLIA (MAMESTRA) OBESULA Sm. Calgary, Alta. ♀
11. ANARTA HAMPA Sm. Mt. Washington, N. H. ♀.
Agrees with type.
12. ANARTA FLANDA Sm. Newfoundland. ♂ Co-type.
13. SCOTOGRAMMA (MAMESTRA) MORANA Sm. Yellowstone Pk., Wyo. ♀ Paratype.
14. SCOTOGRAMMA (TRICHOPOLIA) PTILODONTA Grt. Deming, N. M. ♂.
15. SCOTOGRAMMA (TRICHOPOLIA) PTILODONTA Grt. Deming, N. M. ♀.
These agree with a specimen "compared with type" which is before us.
16. POLIA (MAMESTRA) IMBUNA Sm. Luzern Co., Pa. ♂ Co-type.
17. SCOTOGRAMMA GATEI Sm. Ft. Wingate, N. M. ♂.
18. SCOTOGRAMMA GATEI Sm. Deming, N. M. ♀.
19. POLIA (MAMESTRA) GLACIATA Grt. Palmerlee, Ariz. ♀.
Has been compared with type in Coll. Neumoegen.
20. MIODERA STIGMATA Sm. San Diego, Calif. ♂.
Agrees with co-type in Coll. Barnes.
21. PSEUDOTAMILA VANELLA Grt. Mt. Hood, Oregon. ♂.
Agrees with a specimen before us compared with type in Brit. Museum. It is very probable *vacciniae* Hy. Edw. is this species, but we have not seen the type. In this case Edwards' name would have priority.
22. PSEUDOTAMILA AVEMENSIS Dyar. Awemè, Man. ♂.
23. POLIA (MAMESTRA) LEPIDULA Sm. Redington, Ariz. ♂.
Agrees with a specimen in Coll. Barnes compared with type.

PLATE IV

12

PLATE V

Fig. 1. Syneda ingeniculata Morr. Kerrville, Tex. ♂.
2. Monima (Graphiphora) annulimacula Sm. Kerrville, Tex. ♀.
3. Cerapoda (Oncocnemis) oblita Grt. Reno, Nev. ♂.
4. Cerapoda (Oncocnemis) oblita Grt. Reno, Nev. ♀.
5. Oncocnemis colorado Sm. Manitou, Colo. ♀.
  Agrees with specimen "compared with type" before us.
6. Oncocnemis umbrifascia Sm. Reno, Nev.' ♀.
7. Polia bicolor B. & McD. Kerrville, Tex. ♂ Type.
8. Polia bicolor B. & McD. Kerrville, Tex. ♀ Type.
9. Cleoceris populi Stkr. Calgary, Alta. ♀.
10. Polia palilis Harv. Kerrville, Tex. ♂.
11. Polia palilis Harv. Kerrville, Tex. ♀.
  The sexual difference is considerable in this species.
12. Fotella olivia B. & McD. La Puerta, S. Calif. ♀ Type.
13. Eustrotia bifasciata B. & McD. La Puerta, S. Calif. ♂ Type.
  We fear we have run into the synonymy here; this name will fall before *Yrias albiciliatus* Sm.
14. Phyllophila aleptivoides B. & McD. La Puerta, Cal. ♀ Type.
15. Antaplaga biundulata Zell. Kerrville, Tex. ♂.
16. Pseudanarta crocea Hy. Edw. Esmeralda Co., Nev. ♂.
  Agrees with type in Hy. Edwards' Collection.
17. Graeperia indubitans Wlk.? Kerrville, Tex. ♂.
  This is a new record for our fauna. We are not quite sure of our identification but the specimen agrees well with figure and description as given by Hampson and is structurally a *Graeperia*.
18. Heliolonche indiana Sm. Hessville, Ind. ♂.
  Agrees with co-type in Coll. Barnes.
19. Meliana (Heliophila) ferricola Sm. Tucson, Ariz. ♀.
20. Trichoclea u. scripta Sm. So. Utah. ♂.
21. Trichoclea decepta Grt. So. Ariz. ♂.
22. Trichoclea u. scripta Sm. Colo. ♂.

PLATE V

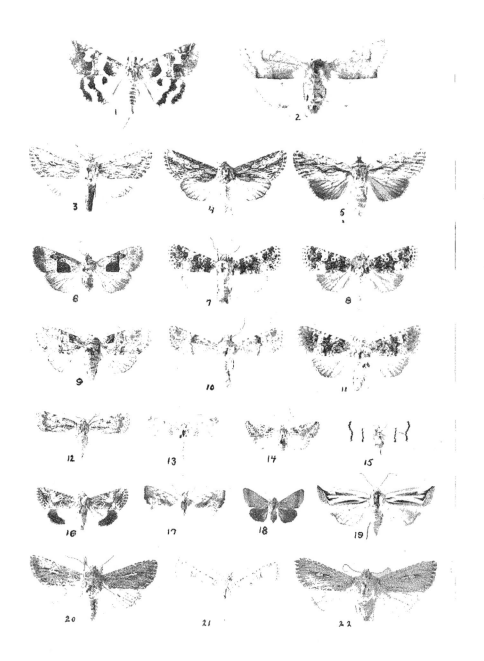

PLATE VI

Fig. 1. POLIA (MAMESTRA) MEODANA Sm. Calgary, Alta. ♀.

2. POLIA DODEI Sm. Calgary, Alta. ♂ Co-type.

3. POLIA LIQUIDA Grt. Vanc. Is. ♂.

4. DISCESTRA (MAMESTRA) CHARTARIA Grt. San Diego, Cal. ♂.

5. DISCESTRA (MAMESTRA) CHARTARIA Grt. San Diego, Cal. ♀.

6. LASYONYCTA OCHRACEA Riley. Alameda Co., Calif. ♂.
One of the original specimens received from Riley himself.

7. TRICHOCLEA EDWARDSI Grt. Solano Co., Calif. ♀.

8. POLIA (SCOTOGRAMMA) MEGAERA Sm. Glenwood Spgs., Colo. ♀ Type.
Hampson makes this a synonym of *densa* Sm. but we doubt the correctness of this action.

9. POLIA DILATATA Sm. Pecos, N. M. ♂.
Agrees exactly with type.

10. POLIA (MAMESTRA) INTENTATA Sm. Redington, Ariz. ♂.
Agrees exactly with type.

11. EPIA ECTRAPELA Sm. Eureka, Utah. ♂.
Agrees with type.

12. ONCOCNEMIS LATICOSTA Dyar. Deming, N. M. ♀.
Agrees with a specimen in our possession from Oslar marked "Co-type."

13. POLIA SARETA Sm. Redington, Ariz. ♂.

14. POLIA SARETA Sm. Redington, Ariz. ♀.
These agree with co-types in Coll. Barnes.

15. POLIA CERVINA Sm. Hymers, Ont. ♂.

16. TRICHOCLEA (MAMESTRA) FUSCOLUTEA Sm. Glenwood Spgs., Colo. ♀.

17. TRICHOPOLIA·(LATHOSEA) URSINA Sm. Glenwood Spgs., Colo. ♂ Co-type.

18. POLIA BASIPLAGA Sm. Palmerlee, Ariz. ♀.
Agrees with co-types in Coll. Barnes.

19. ONCOCNEMIS GRISEICOLLIS Grt. Babaquivera Mts., Ariz. ♀.
Agrees with the type in Neumoegen Coll. Hampson's figure is incorrect.

20. CERAPODA STYLATA Sm. Ft. Wingate, N. M. ♀.
Agrees with type.
POLIA NIPANA Sm. Redington, Ariz. ♂.
The ♂'s are considerably paler in color than the ♀'s.

PLATE VI

PLATE VII

Fig. 1. CIRPHIS (LEUCANIA) CALGARIANA Sm. Calgary, Alta. ♀ Co-type.
2. LEUCANIA RUBRIPALLENS Sm. Glenwood Spgs., Colo. ♂ Type.
3. MELIANA (NELEUCANIA) CITRONELLA Sm. Glenwood Spgs., Colo. ♀ Co-type.
4. LEUCANIA LUTEOPALLENS Sm. (?). ♂ Co-type.
5. CIRPHIS CALPOTA Sm. Harris Co., Tex. ♂ Co-type.
6. MELIANA NIVEICOSTA Sm. Glenwood Spgs., Colo. ♀ Co-type.
7. CIRPHIS MEGADIA Sm. Calgary, Alta. ♀ Co-type.
8. CIRPHIS ANTEROCLARA Sm. Calgary, Alta. ♀ Co-type.
9. AMATHES (ORTHOSIA) AGGRESSA Sm. Denver, Colo. ♂ Co-type.
10. TRICHOCLEA ANTICA Sm. Los. Angeles, Calif. ♂.
The specimen has been compared with the type.
11. SELICANIS CINEREOLA Sm. Glenwood Spgs., Colo. ♂ Co-type.
12. THOLERA (NEURONIA) AMERICANA Sm. Yellowstone Pk., Wyo. ♂.
13. XYLOMANIA (XYLOMIGES) COGNATA Sm. Victoria, B. C. ♂.
14. XYLOMANIA (XYLOMIGES) COGNATA Sm. Calgary, Alta. ♂.
15. XYLOMANIA (XYLOMIGES) COGNATA Sm. South Calif. ♂.
The above three figures represent what we consider to be forms of the one species. The California specimen is more evenly gray and lacks the greenish tinge on primaries characteristic of the other two specimens.
16. XYLOMANIA SIMPLEX Wlk. Glenwood Spgs., Colo. ♂.
The type locality for *simplex* is Vancouver Is., from which locality we have a long series; there is apparently no essential difference between the Colo. and B. C. forms; both are rather variable in maculation and clearness of s. t. line. We have a specimen before us which has been compared with the type of *peritalis* Sm. in the Neumoegen Coll. by Dr. Barnes and marked as agreeing. There is no apparent difference between this and our figured specimen and we imagine the two species will prove identical. The difference in genitalia as shown in Smith's figure may be easily accounted for by the breaking off of one of the arms of the harpe, especially as Smith had but a single specimen before him at the time of description; our specimen agrees in genitalia with Smith's figure of *crucialis* Harv.=*simplex* Wlk.
17. XYLOMANIA PALLIDIOR Sm. Duncans, Vanc. Is., B. C. ♀.
The specimen has been compared and agrees with the type.
18. GRAEPERIA MEGOCULA Sm. Babaquivera Mts., Ariz. ♂.
Agrees with type specimen.
19. STIBADIUM CURIOSUM Neum. Palmerlee, Ariz. ♀.
Agrees with the type in Coll. Neumoegen.
20. ONCOCNEMIS BAKERI Dyar. La Puerta Valley, S. Calif. ♀.
Agrees well with the description of this species; we have not seen the type.

PLATE VII

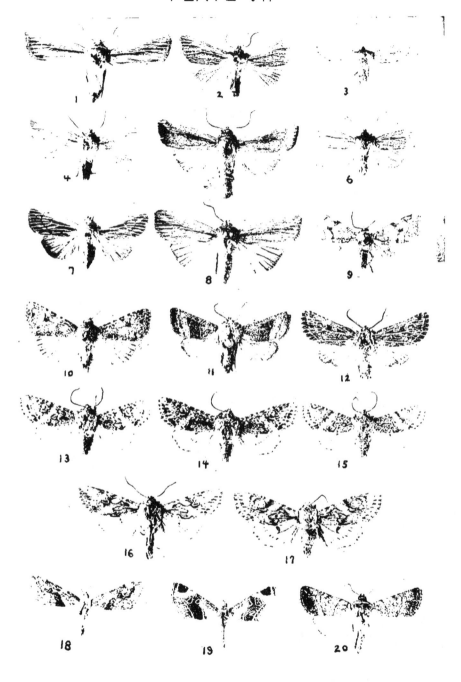

PLATE VIII

Fig. 1.
2.
3.
4.
5. } PERIGRAPHA (GRAPHIPHORA) PRAESES Grt. Duncans, Vanc. Is., B. C.
6.
7.
8.
9.

Figs. 1-3 and 7-9 are ♂'s. This species has a wide range of variation, primaries varying in color from pale brown to deep purple-blackish and having the ordinary spots and terminal space either unicolorous or contrasting. The Figs. 6 and 9 represent ♀ and ♂ of what Smith has described as *G. saleppa;* it appears to us to be but a form of *praeses.*

10.
11.
12. } AMATHES (ORTHOSIA) ANTAPICA Sm. Duncans, Vanc. Is., B. C.
13.

The species varies in color of primaries from gray-brown to deep red-brown, the maculation being clearest in the former specimens.

14. COSMIA (XANTHIA) PULCHELLA Sm. Duncans, Vanc. Is. ♂.
15. COSMIA (XANTHIA) PULCHELLA Sm. Duncans, Vanc. Is. ♀.
16. DRYOTYPE (VALERIA) OPINA Grt. Duncans, Vanc. Is. ♂.
17. DRYOTYPE (VALERIA) OPINA Grt. Duncans, Vanc. Is. ♀.
18. EUROIS (ADELPHAGROTIS) INDETERMINATA Wlk. Duncans, Vanc. Is. ♀.
19. SYNEDA STRETCHII Behr. Pyramid Lake, Nevada. ♂.
20. SYNEDA STRETCHII Behr. Pyramid Lake, Nevada. ♀.

This represents the true *stretchii;* we have compared the specimens with the type in the Strecker Coll. It is listed as a synonym of *howlandi* Grt., but differs from this species in the course of the t. p. line and differently shaped reniform. The secondaries are bright orange-red.

PLATE VIII

PLATE IX

Fig. 1.
2.
3.
4.
5. } XYLOMANIA (STRETCHIA) ERYTHROLITA Grt. San Diego, Calif.
6.
7.
8.

A very variable species. The ground color of primaries may be either red-brown (Fig. 1) or pale purplish gray; specimens of the latter color often show much dark suffusion in subterminal area, culminating in a form as shown at Fig. 6, and which Smith has named *Stretchia acutangula.* The ♀'s are some-times (Fig. 8) almost entirely unicolorous purple-blackish.

9. ANTITYPE UINTARA Sm. San Diego, Calif. ♂.

We imagine this to be the species described by Smith under the above name; we have not seen the type but the specimen agrees well with the description.

10. XYLINA (CALOCAMPA) CURVIMACULA Morr. Duncans, Vanc. Is. ♂.

11. XYLINA MERTENA Sm. Duncans, Vanc. Is. ♂.

Agrees with two ♂ Co-types in Coll. Barnes.

12. GRAPHIPHORA (XYLINA) INNOMINATA Sm. Duncans, Vanc. Is. ♂.

13. GRAPHIPHORA FERREALIS Grt. Duncans, Vanc. Is. ♂.

14. CIRPHIS TEXANA Morr.? Brownsville, Tex. ♂.

We refer this species doubtfully to *texana* as it has considerable re-semblance to *phragmatidicola* and corresponds exactly with the size (29 mm) as given by Morrison. Hampson places *texana* as a synonym of *Borolia ex-tincta* Gn. (*ligata* Grt.) but our present species is certainly not *extincta* as it is generically distinct, falling in the genus *Cirphis* with prothoracic crest and strong anal tuft. The pink suffusion of primaries is wanting, the veins are more clearly strigate and the secondaries are pure white in the ♂ of the species before us. If not *texana* it is probably unnamed, but until we have seen the type of *texana* we prefer to hold the matter in abeyance.

15. XYLOMONIA (XYLOMIGES) MUSTELINA Sm. Pullman, Wash. ♂.

From the type locality and agrees with original description.

16. XYLOMANIA SUBAPICALIS Sm. Duncans, Vanc. Is. ♂.

This is probably a good species and not synonymous with *perlubens* Grt. if Hampson's figure of the latter is at all correct.

17. MONODES NUCICOLORA Gn. Brownsville, Tex. ♀.

18. CIRPHIS SUBPUNCTATA Harv. Brownsville, Tex. ♀.

19. CATABENA ESULA Druce. Brownsville, Tex. ♂.

PLATE IX

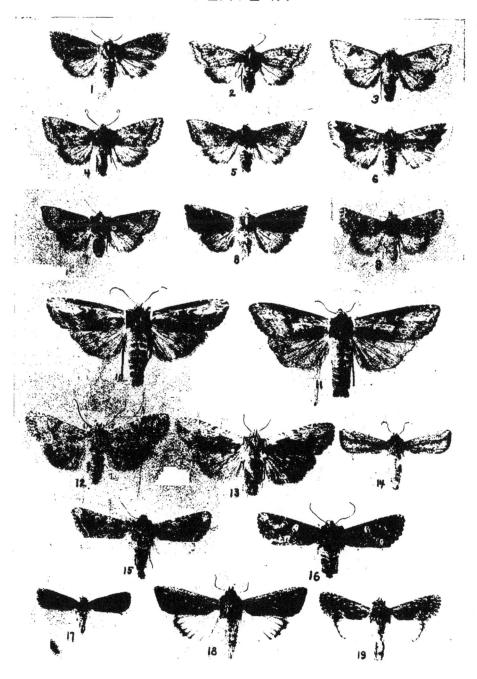

PLATE X

Fig. 1. POLIA (MAMESTRA) TUFA Sm. Eureka, Utah. ♂.
 2. POLIA (MAMESTRA) TUFA Sm. Eureka, Utah. ♀.
 These specimens agree with a ♂ co-type in Coll. Barnes.
 3. POLIA (MAMESTRA) PLICATA Sm. Glenwood Spgs., Colo. ♀.
 Agrees with ♀ co-type in Coll. Barnes.
 4. TRICHOCLEA (MAMESTRA) ARTESTA Sm. Glenwood Spgs., Colo. ♀.
 5. RANCORA BRUCEI Sm. Glenwood Spgs., Colo. ♀.
 The specimen is marked "compared with type" in Smith's handwriting.
 6. COPICUCULLIA LUTEODISCA Sm. Deming, N. M. ♀ Co-type.
 7. CUCULLIA SERRATICORNIS Lint. Vesulia, Calif. ♂.
 Agrees with the type in Albany Museum.
 8. CUCULLIA ASTIGMA Sm. Glenwood Spgs., Colo. ♀
 Agrees with type.
 9. POLIA NEGUSSA Sm. Redvers, Sask. ♀
 10. POLIA TENISCA Sm. Reno, Nev. ♀.
 Agrees well with original description; we have not seen the type.
 11. POLIA ROSEOSUFFUSA Sm. Palmerlee, Ariz. ♂.
 Specimen agrees with type.
 12. POLIA VITTULA Grt. White Mts., Ariz. ♂.
 Agrees with the type in Coll. Neumoegen.
 13. POLIA ULIGINOSA Sm. Redington, Ariz. ♀.
 The specimen agrees with a rather rubbed ♀ co-type in Coll. Barnes; it appears to us to be merely a rather dark form of the preceding.
 14. POLIA NIPANA Sm. Babaquivera Mts., Ariz. ♀.
 Agrees with co-types in Coll. Barnes.

PLATE X

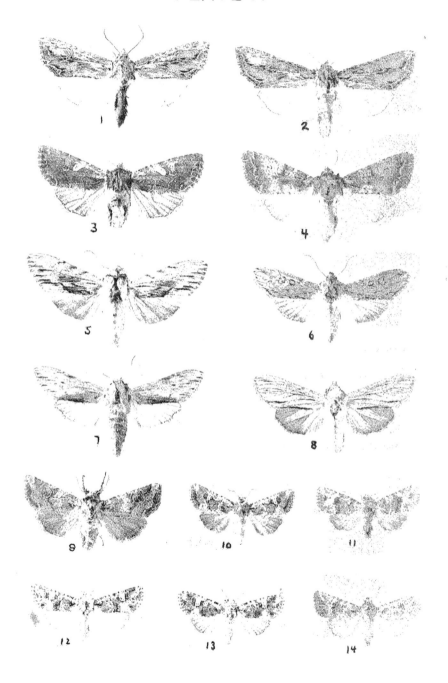

24

PLATE XI

Fig. 1. Psaphidia electilis Morr. Decatur, Ill. ♂.
2. Psaphidia electilis Morr. Decatur, Ill. ♀.
This rare species occurs here the end of March and beginning of April.
3. Amathes (Orthosia) dusca Sm. Cartwright, Man. ♀.
Agrees with ♂ co-type in Coll. Barnes.
4. Eutolype rolandi Grt. New Haven, Conn. ♂.
5. Lathosea pullata Grt. Glenwood Spgs., Colo. ♂.
6. Bryomima (Hadena) chryselectra Grt. White Mts., Ariz. ♂.
7. Homoglaea (Orrhodia) insinuata Sm. Pullman, Wash. ♀.
Agrees with co-type in Coll. Barnes.
8. Hillia algens Grt. Miniota, Man. ♂.
9. Brachylomia (Cleoceris) populi Stkr. Durango, Colo. ♂
10. Homohadena stabilis Sm. Winnipeg, Man. ♂
Agrees with specimen in Coll. Barnes compared with type.
11. Homohadena retroversa Morr. Missouri. ♀.
Agrees with a colored figure of type in the Tepper Collection which we have before us.
12. Homohadena rayata Sm. Shovel Mt., Tex. ♂.
Agrees in the markings of primaries with a ♀ co-type in Coll. Barnes.
13. Homohadena badistriga Grt. Glenwood Spgs., Colo. ♀.
The specimen varies slightly from typical eastern specimens but we imagine is but the local form of this species.
14. Homohadena dinalda Sm. S. Manitoba. ♀.
Probably this species judging from the description; we have not seen the type. It is possibly identical with *fifia* Dyar from Kaslo, B. C., which we do not know.
15. Homohadena (Pericea) loculosa Grt. Yavapai Co., Ariz. ♂.
Agrees with specimen in Coll. Barnes which has been .compared with type in Coll. Neumoegen.
16. Stylopoda anxia Sm. Ft. Wingate, N. M. ♂.
Agrees with co-type in Coll. Barnes.
17. Pseudanarta falcata Grt. Kerrville, Tex. ♀.
Agrees with specimen which has been compared with the type in Coll. Neumoegen.
18. Pseudanarta actura Sm. Wilgus, Ariz. ♀.
Agrees with ♀ co-type in Coll. Barnes.
19. Epidemas cinerea Sm. Glenwood Spgs., Colo. ♂ Type.

PLATE XI

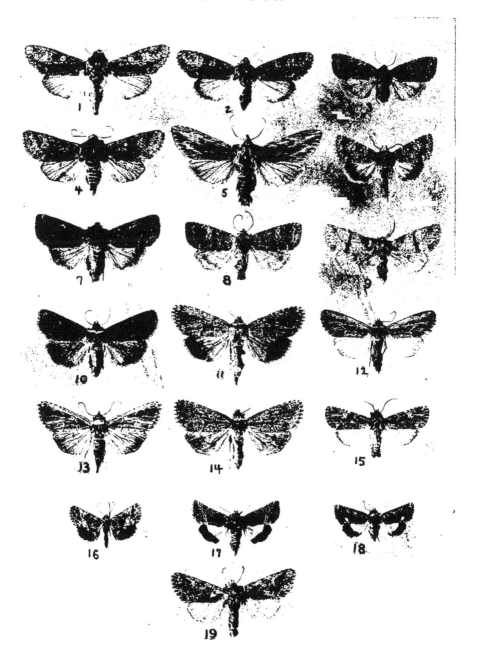

PLATE XII

Fig. 1. XYLOMANIA NICALIS Sm. Pullman, Wash. ♂ Co-type.
2. HOMOGLAEA DIVES Sm. Duncans, Vanc. Is. ♂.
   Agrees with ♀ co-type in Coll. Barnes.
3. GRAPTOLITHA LACEYI B. & McD. Kerrville, Tex. ♂ Type.
4. GRAPTOLITHA (XYLINA) ATINCTA Sm. Cartwright, Man. ♂ Co-type.
5. GRAPTOLITHA TORRIDA Sm. Pullman, Wash. ♂.
   Agrees with two co-types in Coll. Barnes.
6. GRAPTOLITHA HOLOCINEREA Sm. Pullman, Wash. ♂ Co-type.
7. GRAPTOLITHA MERCEDA Sm. Cartwright, Man. ♀ Co-type.
8. GRAPTOLITHA EMARGINATA Sm. Glenwood Spgs., Colo. ♀ Co-type.
   The darker shading apparent in the figure over the whole outer portion of the wings is due to a slight discoloration of the specimen. The color is a pale even gray as seen in Fig. 10.
9. GRAPTOLITHA VIVIDA Dyar. Duncans, Vanc. Is., B. C. ♂.
10. GRAPTOLITHA EMARGINATA Sm. Glenwood Spgs., Colo. ♂ Co-type.
11. EUTOLYPE GRANDIS Sm. Pa. ♀.
12. GRAPTOLITHA LONGIOR Sm. Glenwood Spgs., Colo. ♀ Co-type.

PLATE XII

PLATE XIII

Fig. 1. Litholomia dunbari Harv. Duncans, Vanc. Is. ♂.

2. Acronycta grisea Wlk. Duncans, Vanc. Is. ♀.

We can see no definite point of distinction between this species and *revellata* Sm. of which we possess a co-type. The larval descriptions of the two species given in Hampson's Cat. Lep. Brit. Mus., Vol. VIII are identical also.

3. Trachea (Hadena) divesta Grt. Duncans, Vanc. Is. ♀.

4. Eremobia albertina Hamp. Duncans, Vanc. Is. ♂.

A rather paler form than specimens we have from the type locality, Alberta.

5. Anomogyna (Setagrotis) infimatis Grt. Duncans, Vanc. Is. ♀.

6. Pleroma conserta Grt. Duncans, Vanc. Is. ♂.

7. Xylomania cognata Sm. Duncans, Vanc. Is. ♂.

8. Xylomania himalis Grt. Duncans, Vanc. Is. ♂.

9. Acronycta fragilis Gn. Duncans, Vanc. Is. ♀.

A very large form but apparently identical in markings with the eastern species.

10. Graptolitha ancilla Sm. Duncans, Vanc. Is. ♂.

Agrees with a specimen which has been compared with type. In our opinion there are more names than species in this *georgii* group, but until we have had specimens compared with all the types we prefer to make no definite statements.

11. Graptolitha torrida Sm. Duncans, Vanc. Is. ♀.

12. Scotogramma (Mamestra) yakima Sm. Eureka, Utah. ♂.

Agrees with a specimen identified by Smith.

13. Eurotype (Polia) contadina Sm. Duncans, Vanc. Is. ♀.

Very close to *medialis* Grt. in markings but differs in the structure of the ♂ antennae, they being in this species serrate and not bipectinate.

14. Scotogramma yakima Sm. Eureka, Utah. ♀.

15. Leucocnemis sectiloides B. & McD. Brownsville, Tex. ♂ Type.

The species is very close to *sectilis* Sm. but differs in the browner ground-color of primaries and the presence of slight dashes in the subterminal space.

16. Monima quinquefasciata Sm. Duncans, Vanc. Is. ♂.

17. Leucocnemis sectiloides B. & McD. Brownsville, Tex. ♀.

18.
19. } Monima quinquefasciata Sm. Duncans, Vanc. Is. ♂.

These two figures along with No. 16 represent three varieties of this very variable species or race. They agree with co-types in our possession of Smith's species. No. 16 is pale gray-brown without the reddish tinge present in the other two specimens figured.

PLATE XIII

PLATE XIV

Fig. 1. PARASTICHTIS VERBASCOIDES Gn. Conn. ♂.
 2. PARASTICHTIS VERBASCOIDES Gn. Conn. ♂.
 3. PARASTICHTIS VERBASCOIDES Gn. N. York. ♀
 4. PARASTICHTIS NIGRIOR Sm. Peck, N. J. ♂.
 4. PARASTICHTIS NIGRIOR Sm. Long Is., N. Y. ♀.
 6. PARASTICHTIS NIGRIOR Sm. Long Is., N. Y. ♀.
 7. PARASTICHTIS RORULENTA Sm. Vanc. Is. ♂.

We can see no reason why this should be regarded as distinct from *suffusca* Morr.; we have all manner of intergrades between contrasting specimens and unicolorous ones and present a few of these on this same plate.

 8. PARASTICHTIS VULGARIS G. & R. Decatur, Ill. ♂.

We are not certain about the determinations in this group as we have not had specimens compared with the types, most of which are in the Brit. Museum. The name *vulgaris* G. & R. is, however, usually applied to this species.

 9. PARASTICHTIS CARIOSA Gn. Decatur, Ill. ♀.
 10. PARASTICHTIS VULGARIS G. & R. Elizabeth, N. J. ♀.
 11. PARASTICHTIS CARIOSA Gn. Mo. ♀.

This agrees with a specimen we have in the collection marked *cluna* Stkr.

 12.
 13. } PARASTICHTIS ALIA Gn. Calgary, Alta. ♂'s & ♀.
 14.

This species has gone for years under the name *suffusca* Morr. until Mr. Dod has recently shown (C. Ent. XLII, 192) that *alia* is really the species here figured. Guenee's original description also suits this species better than the one to which the name *alia* has been heretofore applied. Fig. 14 would be according to Smith a typical *suffusca* and 13 might be called *rorulenta*, but as stated above we consider both one species.

PLATE XIV

PLATE XV

Fig. 1.
2.
3. } PARASTICHTIS LIGNICOLORA Gn. Manitoba. 2♂, 2♀.
4.

These specimens are smaller than Eastern *lignicolora* and rather darker in color but apparently similar in maculation; Figs. 2 and 3 have the orbicular smaller and more rounded than is usual in this species but otherwise are identical as far as we can see. It is possible that the name *quaesita* Grt., said by Hampson to be a darker variety of *lignicolora* and described from Wisconsin, will apply to these specimens.

5.
6. } PARASTICHTIS AURANTICOLOR Grt.
7.

This Rocky Mountain form is fairly readily distinguished by the whitish suffusion in the subterminal space; Fig. 5 is a ♂ from Redington, Ariz.; Figs. 6 and 7 ♀'s from Glenwood Spgs., Colo., and White Mts., Ariz., respectively.

8.
9.
10. } PARASTICHTIS LIGNICOLORA Gn.
11.

These are typical Eastern specimens from Mass., Pa., N. J., and Conn.

12.
13. } PARASTICHTIS AURANTICOLOR Grt. Durango, Colo. ♂ and ♀.

Rather smaller specimens with less white suffusion than those figured above.

14. PARASTICHTIS MULTICOLOR Dyar. Victoria, B. C. ♀.

Easily distinguished by the contrasting shading and white edged reniform.

PLATE XV

PLATE XVI

Fig. 1.
   2. } PARASTICHTIS PURPURISSATA B. & McD. Duncans, Vanc. Is., B. C.
   3.   ♂ and ♀ Types and Co-types.
   4.
   5. PARASTICHTIS VULTUOSA Grt. Mass. ♂.
   6. PARASTICHTIS VULTUOSA Grt. Cartwright, Man. ♀.
   7. PARASTICHTIS OCCIDENS Grt. Durango, Colo. ♀.
   8.
   9. } PARASTICHTIS LIGNICOLORA v. ATRICLAVA B. & McD. Duncans, Vanc.
 10.   Is., B. C. ♂ and ♀ Co-types.
 11.
 12. PARASTICHTIS OCCIDENS Grt. Provo, Ut. ♂.
 13. PARASTICHTIS OCCIDENS Grt. Pullman, Wash. ♀.
 14. PARASTICHTIS OCCIDENS Grt. Esmeralda Co., Nev. ♀.

PLATE XVI

PHOTO BY OKO.

PLATE XVII

Fig. 1. TRACHEA BARNESI Sm. Yellowstone Park, Wyo. ♂.
2. TRACHEA BARNESI Sm. Yellowstone Park, Wyo. ♀ Co-type.
3. TRACHEA BARNESI Sm. Yellowstone Park, Wyo. ♀.
Another form of this variable species.
4. TRACHEA ADNIXA Grt. Duncans, Vanc. Is. ♀.
5. TRACHEA ADNIXA Grt. Duncans, Vanc. Is. ♂.
We have identified these specimens as *adnixa* according to a colored figure of one of the types in the Tepper Collection. We have not seen the Brit. Museum specimens.
6. TRACHEA BINOTATA Wlk. Duncans, Vanc. Is. ♂.
7. TRACHEA BINOTATA Wlk. Duncans, Vanc. Is. ♀.
A much darker species than the preceding with the light ochre patch beyond reniform quite prominent.
8. TRACHEA PAVIAE Behr. Palo Alto, Calif. ♂.
Probably a small form of this species.
9. TRACHEA BINOTATA Wlk. Glenwood Spgs., Colo. ♂.
10. TRACHEA BINOTATA Wlk. Glenwood Spgs., Colo. ♀.
A lighter, more clearly marked form of this west coast species; probably a good geographical race.
11. TRACHEA FUMOSA Grt. Glenwood Spgs., Colo. ♂.
Agrees with a specimen in the Schaus Coll. which is marked "compared with type."
12. TRACHEA PAVIAE Behr. San Francisco, Calif. ♂.
13. TRACHEA PAVIAE Behr. California. ♀.
Agrees with a specimen which we have compared with the type in the Strecker Collection. The species is rather a uniform brown with a large yellowish patch beyond the reniform.
14. TRACHEA FUMOSA Grt. Glenwood Spgs., Colo. ♀.
15. TRACHEA FINITIMA Gn. Glenwood Spgs., Colo. ♂.
16. TRACHEA FINITIMA Gn. Maine. ♀.
17. TRACHEA CERIVANA Sm. Cartwright, Man. ♂.
Agrees with specimens so named by Smith.
18. TRACHEA REMISSA Hbn. Liberty, N. Y. ♂.
19. TRACHEA CERIVANA Sm. Duncans, Vanc. Is. ♂.
A specimen with the median area more heavily brown than in Fig. 17.

PLATE XVII

PLATE XVIII

Fig. 1. TRACHEA DELICATA Grt. Illinois. ♀.
2. TRACHEA SMARAGDINA Neum. Santa Catalina Mts., Ariz. ♂.
3. TRACHEA SMARAGDINA Neum. Redington, Ariz. ♀.
4. TRACHEA MARINA Grt. San Diego, Calif. ♂.
5. TRACHEA MARINA Grt. San Diego, Calif. ♂.
6. TRACHEA VIRIDIMUSCA Sm. Oconee, Ill. ♂.
7. TRACHEA MISELOIDES Gn. Oconee, Ill. ♂.
8. TRACHEA MISELOIDES Gn. Decatur, Ill. ♀.
9. TRACHEA INORDINATA Morr. N. Hamp. ♂.
10. TRACHEA MONTANA Sm. Denver, Colo. ♂.
11. TRACHEA MONTANA Sm. Denver, Colo. ♂.
A better marked variety of the preceding.
12.
13.
14.
15. } TRACHEA CINEFACTA Grt.
16.
17.
18.

Fig. 12 is a ♂ from Red Deer River, Alta.; Figs. 13 and 14 a pair from Duncans, Vanc. Is.; Figs. 15 and 16 a pair from San Diego, Calif., and Figs. 17 and 18 a pair from Reno, Nevada. Fig. 16 is closest to the type in the Hy. Edwards Coll.

19. TRACHEA CENTRALIS Sm. Glenwood Spgs., Colo. ♂.
20. TRACHEA CENTRALIS Sm. Glenwood Spgs., Colo. ♀.

We are doubtful as to the identity of these specimens; they are close to *centralis* from the Sierra Nevada Mts., Calif., but there are certain points of difference which may prove that the Colorado species is either distinct or a good race when we secure more material.

PLATE XVIII

PLATE XIX

This plate represents a series of specimens from various localities which have received different names but which we consider will probably prove to be but one variable species. Fig. 1 is a ♂ of what is known as *Andropolia* (*Polia*) *diversilineata* Grt. from Arizona; Fig. 2 a ♂ from Plumas Co., Calif., representing a great reduction of maculation and a paling of the ground color; this is *A. resoluta* Sm. Figs. 3 and 4 are ♀'s from Colorado of *diversilineata;* Figs. 5-7 ♀'s from Provo, Utah, showing the gradual change from a fairly uniform dark gray form (Fig. 6) to a checkered black and white form (Fig. 7) which Smith has recently named *submissa;* Figs. 8 and 9 are dark gray ♀'s from Yavapai Co., Ariz., corresponding to the ♂ in Fig. 1. In the following plate we give further illustrations of these forms.

PLATE XIX

PLATE XX

A continuation of the preceding plate; Figs. 1-5 are ♂'s from Provo, Utah, showing a gradual increase in depth of ground color and markings; Fig. 2 is a fairly typical *illepida* Grt., whereas Fig. 5 can be referred to *submissa* Sm. Fig. 6 is a ♂ from Durango, Colo., and Fig. 7 a specimen of the same sex from Ft. Wingate, N. M., approaching the dark Arizona form; Figs. 8-12 represent ♀'s from Provo, Utah, showing both pale, rather immaculate forms (*illepida*) and dark well marked specimens (*diversilineata*); Fig. 13 is a ♀ from Glenwood Spgs., and Fig. 14 one from Ft. Wingate, N. M. In S. Colo., New Mexico and especially in Arizona *diversilineata* appears to have developed into a good geographical race, at least all our Arizona specimens are constant in this respect. In Utah the *illepida* form predominates, but there are occasional specimens of *diversilineata* found, especially among the ♀'s, and these are apparently diverging into a new form, *submissa* Sm., which can, however, scarcely be considered more than an aberration at present. *Resoluta* Sm. is nothing but a slightly paler *illepida;* we have a specimen which has been compared with the ♂ type in the Nat. Museum.

PLATE XX

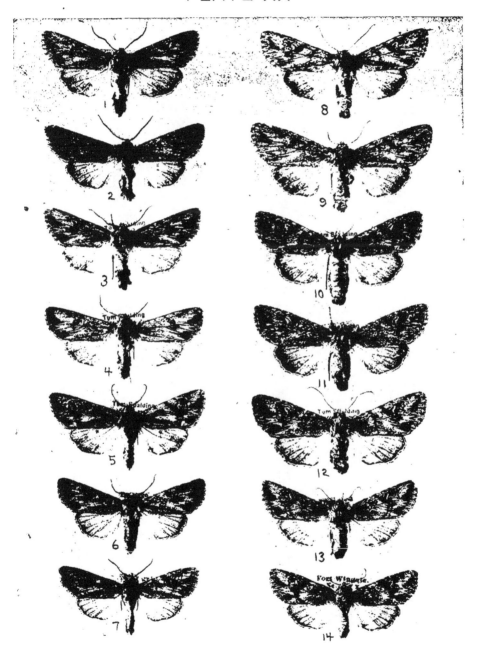

44

PLATE XXI

Fig. 1. TARACHE BEHRI Sm. Gila Co., Ariz. ♀.
The specimen agrees with one before us that has been compared with the type.

2. TARACHE ANILUNA Sm. Babaquivera Mts., Ariz. ♀ Co-type.

3. CONACONTIA HUACHUCA Sm. Huachuca Mts., Ariz. ♂ Co-type.

4. TARACHIDIA BINOCULA Grt. Cartwright, Man. ♂.
The type locality is Texas but our specimen agrees well with Hampson's figure of the type (Cat. Lep. Brit. Mus., Pl. 170, Fig. 3).

5. TARACHE PIMA Sm. Babaquivera Mts., Ariz. ♂ Co-type.

6. TARACHE DISCONNECTA Sm. Huachuca Mts., Ariz. ♂ Co-type. ·

7. TARACHE EXPOLITA Grt. Santa Catalina Mts., Ariz. ♀.
Agrees with specimen before us that has been compared with the type in Coll. Neumoegen.

8. CONOCHARES ALTERA Sm. Deming, New Mexico. ♀.
Has been compared with the type specimen.

9. TARACHE SEDATA Hy. Edw. Deming, New Mexico. ♂.
The type has the dark costal spot joined to the basal patch but otherwise our specimen is identical. Apparently *niveicollis* Sm. is a synonym of this species according to a specimen in our possession marked "compared with type" by Smith; *nuicola* Sm., placed by Hampson as a synonym, is not this species but belongs in the genus *Graeperia* and is probably the same species as that renamed *carcharodonta* by Hampson (Cat. Lep. Brit. Mus., 10, 677)

10. CONOCHARES ACUTUS Sm. Redington, Ariz. ♂.
Agrees with the ♂ type in our possession.

11. CONOCHARES ACUTUS Sm. Redington, Ariz. ♀.

12. CONOCHARES INTERRUPTUS Sm. Walters Sta., Calif. ♀ Co-type.
Probably the same species as *arizonae* Hy. Edw.

13. CONOCHARES CATALINA Sm. Santa Catalina Mts., Ariz. ♀.
Agrees with co-type in Coll. Barnes. Doubtless a pale form of *acutus* Sm.; we have a long series with all manner of intergrades.

14. TARACHIDIA HUITA Sm. S. Ariz. ♀ Co-type.

15. TARACHIDIA CUTA Sm. Gila Co., Ariz. ♂.
Agrees with co-type in our possession.

16. CONOCHARES HUTSONI Sm. Colo. Desert, Yuma Co., Ariz. ♀ Co-type.

17. TARACHE SEMIATRA Sm. Yuma Co., Ariz. ♂.
Agrees with co-type in our possession.

18. TARACHE DIMIDIATA Sm. So. Ariz. ♂.
Agrees with ♀ type in our possession.

19. TARACHIDIA (SPRAGUEIA) FUMATA Sm. Verdi, Nev.
Agrees with co-type in Coll. Barnes.

20. CHAMAECLEA FERNANA Grt. Redington, Ariz. ♀.
Agrees with a specimen in our possession that has been compared with the type in Coll. Neumoegen.

PLATE XXI

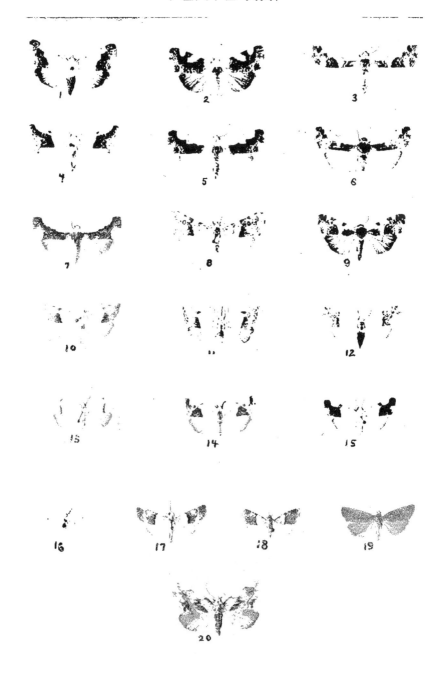

INDEX

	Page
actura Sm.	44
acutus Sm.	44
adnixa Grt.	36
aggressa Sm.	16
albertina Hamp.	28
albiciliatus Sm.	12
aleptivoides B. & McD	12
algens Grt	24
alia Gn.	30
alma Stkr.	4
altera Sm.	44
americana Sm.	16
ancilla Sm.	28
aniluna Sm.	44
annulimacula Sm.	12
antapica Sm.	18
anteroclara Sm.	16
antica Sm.	16
anxia Sm.	24
artesta Sm.	22
ascula Sm.	6
astigma Sm.	22
atincta Sm.	26
v. atriclava B. & McD	34
auranticolor Grt.	32
avemensis Dyar	10
badistriga Grt.	24
bakeri Dyar	16
barnesi Sm.	36
basiplaga Sm.	14
behri Sm.	44
bicolor B. & McD	12
bifasciata B. & McD	12
biforma Sm.	10
binocula Grt.	44
binotata Wlk.	36
biundulata Zell.	12
brucei Sm.	22
buta Sm.	6
calgariana Sm.	16
calpota Sm.	16
carcharodonta Hamp.	44
cariosa Gn.	30
catalina Sm.	44
centralis Sm.	38
cerivana Sm.	36
cervina Sm.	14
chartaria Grt.	14
chryselectra Grt.	24
cinefacta Grt.	38
cinerea Sm.	24
cinereola Sm.	16
circumcincta Sm.	8
citronella Sm.	16
cognata Sm.	16, 28
connecta Grt.	6
conserta Grt.	28
contadina Sm.	28
crassipes Sm.	10
cretata G. & R.	6
crocea H. Edw.	12
crucialis Harv.	16
curiosum Neum.	16
curvimacula Morr.	20
cuta Sm.	44
decepta Grt.	12
decolor Morr.	8
delicata Grt.	38
dilatata Sm.	14
dimidiata Sm.	44
dinalda Sm.	24
disconnecta Sm.	44
diversilineata Grt.	40
dives Sm.	26
divesta Grt.	28
dodei Sm.	14
dunbari Harv.	28
dusca Sm.	24
ectrapela Sm.	14
edwardsi Grt.	14
electilis Morr.	24
emarginata Sm.	26
erratica B. & McD	6
erythrolita Grt.	20

	Page		Page
esula Druce	20	longior Sm.	26
expolita Grt.	44	luteodisca Sm.	22
falcata Grt.	24	luteopallens Sm.	16
ferrealis Grt.	20	marina Grt.	38
ferricola Sm.	12	megadia Sm.	16
finitima Gn.	36	megaera Sm.	14
flanda Sm.	10	megocula Sm.	16
fragilis Gn.	28	meodana Sm.	14
fulvia Edw.	4	merceda Sm.	26
fumata Sm.	44	minuscula B. & McD.	6
fumosa Grt.	36	misellus Sm.	6
fuscolutea Sm.	14	miseloides Gn.	38
gatei Sm.	10	montana Sm.	38
glaciata Grt.	10	morana Sm.	10
grandis Sm.	26	multicolor Dyar	32
grisea Wlk.	28	mustelina Sm.	20
griscicollis Sm.	14	negussa Sm.	22
hampa Sm.	10	nicalis Sm.	28
hanhami Sm.	8	nigrior Sm.	30
himalis Grt.	28	nipana Sm.	22
hirsutaria B. & McD.	8	nivalis Sm.	6
holocinerea Sm.	26	niveicollis Sm.	44
howlandi Grt.	18	niveicosta Sm.	16
huachuca Sm.	44	noverca Grt.	8
huita Sm.	44	nucicolora Gn.	20
hutsoni Sm.	44	nuicola Sm.	44
hydromeli Harv.	6	obesula Sm.	10
illepida Grt.	42	oblita Grt.	12
imbuna Sm.	10	occidens Grt.	34
indeterminata Wlk.	18	ochracea Riley	14
indiana Sm.	12	oculata Sm.	10
indubitans Wlk.	12	olivia B. & McD.	12
infimatis Grt.	28	opina Grt.	18
ingeniculata Morr.	12	orida Sm.	6
innominata Sm.	20	ortruda Sm.	10
inordinata Morr.	38	palilis Harv.	12
insinuata Sm.	24	pallidior Sm.	16
intentata Sm.	14	paviae Behr.	36
interruptus Sm.	44	peritalis Sm.	16
jaguarina Gn.	10	perlubens Grt.	20
laceyi B. & McD.	26	pernana Grt.	44
laticosta Dyar	14	petulans H. Edw.	10
lepidula Sm.	10	pima Sm.	44
lepetita Sm.	6	plicata Sm.	22
lignicolora Gn.	32	populi Stkr.	12, 24
liquida Grt.	12	praeses Grt.	18
loda Stkr.	8	ptilodonta Grt.	10
loculosa Grt.	24	pulchella Sm.	18
longiciliata Hlst.	6	pullata Grt.	24

	Page
purpurissata B. & McD.........	34
quaesita Grt.	32
quinquefasciata Sm.	28
rarata Wlk.	6
rayata Sm.	24
remissa Hbn.	36
reniformis Grt.	8
reniformis Sm.	10
resoluta Sm. 40, 42
rolandi Grt.	24
rorulenta Sm.	30
roseisuffusa Sm.	22
rubripallens Sm.	16
sareta Sm.	14
sectiloides B. & McD..........	28
sedata H. Edw.	44
semiatra Sm...................	44
serraticornis Lint.	22
simplex Sm.	16
smaragdina Neum.	38
snowi Sm.	6
stabilis Sm.	24
v. stephensi Skin.	4
stigmata Sm.	10
stretchii Behr.	18
strigilis Grt.	8
stylata Sm.	14

	Page
subapicalis Sm.	20
submissa Sm. 40, 42
subpunctata Harv.	20
subrufa B. & McD............	6
suffusca Morr.	30
tenisca Sm.	22
texana Morr.	20
torrida Sm. 26, 28
transparens Grt.	8
tristigmata Grt.	8
tuana Sm.	8
tufa Sm.	22
uintara Sm.	20
uliginosa Sm.	22
umbrifascia Sm.	12
ursina Sm.	14
u. scripta Sm.	12
vacciniae H. Edw.	10
vanella Grt.	10
verbascoides Gn.	30
viridimusca Sm.	38
vittula Grt.	22
vivida Dyar	26
vulgaris G. & R..............	30
vultuosa Grt.	34
yakima Sm.	28

TO THE

NATURAL HISTORY

OF THE

LEPIDOPTERA

OF

NORTH AMERICA

VOL. II
No. 2

THE N. AMERICAN SPECIES
of
THE LIPARID GENUS OLENE

BY

WILLIAM BARNES, S. B., M. D.

AND

J. H. McDUNNOUGH, Ph. D.

DECATUR, ILL.
THE REVIEW PRESS
APRIL 15, 1913

Published
Under the Patronage
of
Miss Jessie D. Gillett
Elkhart, Ill.

INTRODUCTION

The species of the Liparid genus *Olene* Hbn. (*Parorgyia* Pack.) have proved a great stumbling block to all N. American collectors and compilers of lists. This is partly due to several very incomplete and inadequate descriptions by some of the older authors, which have resulted in constant misidentification, almost each worker on the group arriving at a different conclusion regarding the identity of certain species from that which previously existed. A lack of knowledge of the early stages has also added to the confusion, combined with a difficulty in properly associating the two sexes of a species, due to considerable sex-dimorphism. These latter difficulties are being slowly overcome, but the general apathy of collectors in N. America towards publishing larval notes—a state of affairs which has existed for the past ten years—has rendered progress in this direction much slower than necessary. Our studies of the material received from various collectors prove conclusively that species of which nothing definite is known of the earlier stages have been bred on several occasions. As the species of this genus are on the whole much more readily separated in the larval than in the adult stage, what a boon a few notes on the larvae would have been to us at the present moment!

Taking Dyar's list (Bull. 52, U. S. N. Mus.) as a basis and adding those species described at a later date we find we have eighteen names at present included in the genus, either as good species, varieties or synonyms. Of these *tephra* Hbn. (1805), *plagiata* Wlk. (1855), and *atomaria* Wlk. (1856) were described without locality; the first undoubted N. Am. species are *achatina* and *leucophaea*, figured by Abbot and Smith, together with larvae, in Lep. Ins. Ga. II, Pl. 77 and 78 (1797). In 1864 Packard described *basiflava* and in 1866 Grote and Robinson added four more species, *clintonii, cinnamomea, obliquata* and *parallela*. These were followed by *atrivenosa* Palm (1893), *manto* Strecker (1900) and *montana* Beutenmuller (1903). In 1911 Dyar added var. *interposita, pini,* var. *pinicola* and *grisefacta*, which were followed in the same year by *styx* Barnes & McDunnough.

Numerous attempts at a correct tabulation of these various names have been made.

Riley in 1887 doubtfully suggests the following synonymy (Proc. Ent. Soc. Wash. I, 88).

leucophaea A. & S.
    obliquata G. & R.?
achatina A. & S.
    clintonii G. & R.
    parallela G. & R.?
    basiflava Pack.?
    cinnamomea G. & R.?

Packard in 1890 (5th Rep. of the U. S. Ent. Comm. 135-137) alters this to the following:

leucophaea A. & S.
    clintonii G. & R.
    basiflava Pack.?
    obliquata G. & R.?
    cinnamomea G. & R.?
achatina A. & S.
    parallela G. & R.?

Dyar in 1894 (Psyche VII, 135-7) on the strength of breeding experiments by himself and Seifert offers the following list:

leucophaea A. & S.
    clintonii G. & R.
  var. bäsiflava Pack.
achatina A. & S.
    parallela G. & R.
  var. obliquata G. & R.
cinnamomeä G. & R.
plagiata Wlk.

Neumoegen & Dyar in their revision of N. Am. Bombyces (Jour. N. Y. Ent. Soc. II, 57 (1894) for the first time introduce the generic term *Olene* Hbn. and list the species as follows:

cinnamomea G. & R.
achatina A. & S.
    parallela G. & R.
  var. tephra Hbn.
    obliquata G. & R.
leucophaea A. & S.
    clintonii G. & R.
  var. basiflava Pack.
  var. atrivenosa Palm.
plagiata Wlk.
    atomaria Wlk.

Beutenmüller in 1897 (Bull. Am. Mus. N. Hist. IX, 210) from information regarding Walker's types received from Sir Geo. Hampson gives the following synonymy for *plagiata* Wlk.

plagiata Wlk.
   atomaria Wlk.
   clintonii G. & R.

Dyar's list of N. Am. Lepidoptera (Bull. 52, U. S. N. M. p. 260, 1892) adopts the following synonymy:

achatina A. & S.
   parallela G. & R.
   var. tephra Hbn.
      obliquata G. & R.
   var. cinnamomea G. & R.
leucophaea A. & S.
   var. basiflava Pack.
   var. atrivenosa Palm.
   var. manto Stkr.
plagiata Wlk.
   atomaria Wlk.
   clintonii G. & R.

This arrangement stood without further alterations until 1911 when Dyar (Proc. Ent. Soc. Wash. XIII, 16) published a very useful paper on the species under this genus, dividing them into two groups, according as the larvae were known to feed on pine or deciduous trees; he made no complete attempt at synonymy, but emphasized the fact that there were probably more good species present in N. America than had been recognized in his List of 1902.

Our own studies of the group lead us to concur with this last statement of Dyar, and we are further of the opinion that several of the older species have been misidentified up to the present time, and that others must be omitted altogether from our list. In this latter category we include *tephra* Hbn. (Samml. Ex. Schm. I, Pl. 178) and *plagiata* Wlk. (1855 Cat. Brit. Mus. IV, 799). The figure of *tephra* is totally unlike any N. American species we have ever seen, the difference being especially marked in the secondaries of the ♂, which show in the figure a distinct waved black postmedian line followed by a dark terminal shade, and strongly checkered fringes, features unknown to us in N. Am. species of *Olene*. Either the figure must be regarded as very poor, or else the specimens must be taken to be non-American. We prefer the latter alternative as in the first case an absolute identification is impossible and far too much scope is given to the imagination

50

of the investigator. Regarding *plagiata* Wlk. Mr. Wolley Dod, who has recently visited the British Museum and has seen the type, surprised us with the statement that this was nothing but *leucostigma* A. & S. Sir Geo. Hampson, in answer to our inquiries, confirmed Mr. Dod's opinion and further sent us a colored figure and a photograph of the type which removes all doubt from the question; we reproduce a figure of same (Pl. 7, Fig. 1).

A species described by Walker as *Edema plagiata* (1865 Cat. Brit. Mus. 32, 427) and at present listed by Dyar under *Symmerista* (No. 3125. 1) should by rights fall into the genus *Olene*. As far back as 1868 Grote & Robinson, in their remarks on Walker's types, recognized this (Tr. Am. Ent. Soc. II, 86), but their statement has been ignored by subsequent authors. Neumoegen & Dyar state (Jn. N. Y. Ent. Soc. II, 173, 1894) that the type is lost, but this is erroneous as we have a colored figure and a photograph of it received through the kindness of Sir. Geo. Hampson. It is a pity that the name *plagiata* will have to be introduced into the genus for a different species from that to which until now has been applied, as some confusion is liable to occur, but the name appears perfectly eligibly, the old *plagiata* falling into a different genus.

We have examined in the course of our studies the male genitalia of nearly every species but find that they are useless as a means of differentiation; they are practically identical throughout the whole

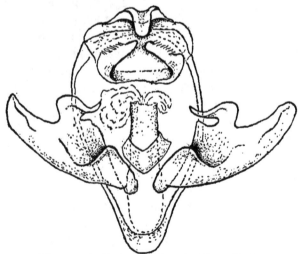

Fig. 1. ♂ Genitalia of o. *basiflava* Pack.

group and any slight differences that may occur could as well be ascribed to individual variation as to specific distinction. As we before remarked the most easily recognized points of differentiation between the species probably will be found to occur in the larvae and until these have been carefully studied and extensive breeding ex ova undertaken, all attempts at a thoroughly reliable classification must be more or less tentative; it is very probable that several alterations in the following arrangement will have to be made before stability is attained, but such things are unfortunately unavoidable in a group so imperfectly known as this one.

Besides the material in Collections Barnes and Merrick (recently acquired by Dr. Barnes) we have had before us, with the exception of types, all the material from the American Museum, New York; Brooklyn Ins. Museum; Field Museum, Chicago; Central Experiment Farm, Ottawa, Canada; Collections Chagnon, Cockle, Doll, Lacey and Winn; further through the kindness of Messrs. Grossbeck, Gerhard, Hampson, Johnson and Palm we have been able to secure photographs and valuable information concerning the types of species described by Grote and Robinson, Strecker, Walker, Palm and Packard. To all these gentlemen and institutions we would express our sincere obligation; without their kind assistance this work would have been an impossibility. We regret deeply that owing to matters of a personal nature the Curator of the National Museum, Washington, has seen fit to refuse us any assistance either in the way of photographs or specimens; we had hoped the time had arrived when personal bickerings would be laid aside in the cause of science, but apparently we were too optimistic.

SYSTEMATIC PORTION

GENUS OLENE Hbn. (Type, mendosa Hbn.)

Olene Hubner, Zutr. Exot. Schmett. II, 19. (1823); Moore, Lep. Ceylon, II, 95
(1883); Hampson, Moths Brit. Ind. I, 452 (1892); Neumoegen &
Dyar, Jour. N. Y. Ent. Soc. II, 30 (1894).
Parorgyia Packard, Proc. Ent. Soc. Phil. III, 332 (1864).

We have been unable to secure any specimens of the type species
of *Olene,* which occurs in Java, for examination; if *mendosa* should
prove generically distinct from our N. Am. species, then these latter
will fall under Packard's genus *Parorgyia.* Packard specified no type
when erecting this genus; as first species he lists *"achatina* Hbn." but
the species he really had before him, as we have verified by photographs
of the specimens in the Harris Coll. mentioned by him under this
heading, was *atomaria* Wlk. (*obliquata* G. & R.). The second species
mentioned, *"leucophaea* A. & S." was evidently unknown to him; we
believe therefore the safest course would be to consider *basiflava* Pack.,
the 3rd species listed, genotype. All three species are in any case con-
generic, so that no confusion is liable to ensue.

We offer the following generic definition, based on N. Am.
material.

Antennae bipectinate, very lengthily in ♂, shortly in ♀; eyes naked, round;
palpi porrect, extending beyond front, heavily clothed with hair beneath, almost
concealing 3rd joint, which is short; front smooth; thorax clothed rather roughly
with hair and hair-like scales, metathorax with prominent divided tuft of metallic scales; abdomen with button-like tufts of metallic scales on 2nd and 3rd seg-

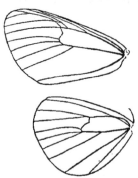

Fig. 2. Venation of Genus *Olene* Hbn.

ments dorsally; tibiae unspined, clothed with long hair; primaries with costa
well rounded at apex, outer margin convex, 12 veined, areole present, vein R_1
from beyond middle of cell; R_2 from areole; R_3 and R_4 on long stalk from apex
of areole; R_5 connate with R_3 and R_4 or else very slightly stalked with same;

M₁ from lower angle of areole; M₂, M₃ and Cu₁, from around lower angle of cell; Cu₂ from well before angle, discocellulars obsolete. Secondaries well rounded outwardly, S. C. separating widely from cell before middle of same. discocellulars obsolescent, angled inwardly; R₁ and M₁ shortly stalked; M₂ from just above lower angle of cell; M₃ and Cu₁, connate from lower angle of cell or slightly stalked; Cu₂ from well before lower angle; frenulum present.

The areole varies considerably in size, specimens with a reduced areole usually have vein R₅ slightly stalked with R₃ + R₄; veins M₃ and Cu₁ of secondaries are also variable in position.

O. ACHATINA A. & S.

Phalaena achatina Abbot & Smith, Lep. Ins. Ga. II, 153, Pl. 77 (1797).
Orgyja achatina Hubner, Verz. bek. Schmett, 161, 1677 (1816).
Dasychira achatina Walker, Cat. Lep. Brit. Mus. IV, 865 (1855); Morris, Syn.
    Lep. N. Am. p. 257 (1862).
Parorgyia achatina Grote & Robinson, Proc. Ent. Soc. Phil. VI, 3 (1866); Pack-
    ard, 5th Rep. U. S. Ent. Comm. p. 135 (1890).
Olene achatina Dyar, Bull. 52, U. S. N. Mus. p. 260 (1902); id. Proc. Ent. Soc.
    Wash. XIII, 16 (1911).

This species has never been definitely identified; it has been largely confused with *parallela*, G. & R., which is a form of *obliquata* G. & R., by Packard and late authors, but, as Dyar points, out the larva of this latter form, as described by Seifert, differs greatly from Abbot's figure. The larva, as depicted by Abbot, approaches very closely that of the Florida form of *basiflava* Pack., which has generally been confused with *leucophaea* A. & S. for some reason or other. These larvae, how-ever, have scattered black plumed hairs in the lateral tuftings, which are entirely absent in Abbot's figures. Apart from the prominent logi-tudinal dash the figures of the adults bear out the relationship above hinted at; both the Florida form and the true *basiflava* Pack. *(clintonii* G. & R.) have that mossy appearance in the fresh ♂'s which led Abbot to bestow the name *achatina;* furthermore all three forms show promi-nent black dashes in the brown subterminal band of the ♀ near apex of primaries, a feature not found in *obliquata* G. & R. or *parallela* G. & R.

The black dash in submedian fold, which Abbot mentions as the distinguishing character of *achatina* may or may not be of specific value; Seifert has proved conclusively that in *obliquata* G. & R. it may be present or absent. In the material before us of the Florida form we have found no specimens with such a dash, which of course does not prove that it may not occur; it is quite possible that Abbot's figure of the larva is inaccurate and that what we call the Florida form of *basiflava* may prove to be merely a form of *achatina* with the dash absent; until however this can be definitely proved by breeding we consider it wiser to treat them as separate species.

HABITAT. Georgia (A. & S.).

O. BASIFLAVA Pack. (Pl. I, Figs. 1-6; Pl. 6, Fig. 7).

Parorgyia basiflava Pack. Proc. Ent. Soc. Phil. III, 333 (1864).
Parorgyia clintonii Grote & Robinson Proc. Ent. Soc. Phil. VI, 3, Pl. I. figs. 2 and 3 (1866).

We have received a photograph of Packard's type (a ♂ from the Harris Coll. taken at Nonantum, a suburb of Boston, Mass.) through the kindness of Mr. Johnson, Curator of the Bost. Soc. Nat. Hist.; unfortunately the condition is such that a reproduction has been impossible. We have also a photograph of Grote & Robinson's ♂ type of *clintonii* (Pl. 6, Fig. 7), from Rhode Island, received from Am. Mus. Nat. Hist. and consider it identical with *basiflava* Pack. Packard's description was evidently based on a faded, worn specimen, but he lays stress on the fact that the base of primaries is yellow with costal margin dark cinereous. Grote & Robinson dismiss the ♂ in a few words as similar to the ♀ but smaller, with less brown and markings more indistinct; from the ♀ description we note however that "the space between basal line and the inner median line is pale brown, cinereous olivaceous along costa." The photograph of the ♂ type shows a pale basal area just as in *basiflava,* and further the characteristic dark dashes in the subterminal space; the type is labelled "Grote & Robinson Collection No. 23087." The ♀ type of *clintonii* has no type label, but among the material sent from New York for study purposes, we have discovered a ♀ specimen labelled "Grote & Robinson Coll. No. 23088" (Pl. I, Fig 4) which so exactly corresponds with the figure of the ♀ type given with the original description that we have no hesitation in declaring it to be the missing type, and have marked it as such. Early authors seem to have associated *clintonii* with a form of *basiflava* without the yellow basal area, but this is not borne out by the photograph of the type. Grote at various times refers to *clintonii* as a valid species but persistently ignores *basiflava* Pack. The fact that *clintonii* G. & R. has latterly been associated with *plagiata* Wlk. is due largely to the fact that the specimens in the Grote Coll. in the British Museum labelled *clintonii* are not the true species. We have received a colored figure of one of the British Museum *clintonii* and believe it to be *obliquata* G. & R.; it lacks the pale basal space and the subterminal dashes which characterize true *clintonii*. Sir Geo. Hampson was correct in associating the Brit. Museum species with *atomaria* Wlk. but erred in stating to Beutenmueller regarding *plagiata* Wlk. that "as far as the condition of the types allows me to judge, this is the same as *Parorgyia clintonii* G. & R." (Bull. Am. Mus. N. Hist. IX, 210, 1897). *Plagiata* Wlk. must now be considered identical with *leucostigma* A. & S. and the true *clintonii* was evidently unknown to Hampson. Beutenmueller's synonymy, as published in the above reference, has been followed in Dyar's list without any further attempt on the latter author's part to verify the truth of this statement. Riley, who doubtfully placed *clintonii* as a synonym of *achatina* A. & S., came very much nearer the mark, and it is still quite possible that the two may be but forms of a single species.

Packard's reference of this species to *leucophaea* A. & S. is due to an entire misidentification; the larvae are totally different.

We recognize two subspecies or geographical races, the typical *basiflava basiflava* occurring in the New England States, New York, Pennsylvania and as far south as Maryland, whilst a second form, which we have referred to under *achatina* as the "Florida form", occurs in Florida and the Southern Atlantic States as far north as N. Carolina; we characterize this below.

(a) O. BASIFLAVA BASIFLAVA Pack. (Pl. I, Figs. 1, 2).

Parorgyia basiflava Packard, Proc. Ent. Soc. Phil. III, 333 (1864).
Parorgyia clintonii Grote & Robinson, Proc. Ent. Soc. Phil. VI, 3, Pl. I, figs. 2, 3 (1886); Riley, Proc. Ent. Soc. Wash. I, 88, 1887 (*achatina?*); Grote, Can. Ent. XIX, 113-114 (1887).
Parorgyia leucophaea Packard (nec Abbot. & Sm.); 5th Rep. U. S. Ent. Comm. 138 (1890).
Parorgyia leucophaea v. *basiflava* Dyar, Psyche VII, 135 (1894) (larva).
Olene leucophaea v. *basiflava* Neumoegen & Dyar, Jour. N. Y. Ent. Soc. II. 58 (1894); Dyar, Bull. 52, U. S. Nat. Mus. 260 (1902).
Dasychira plagiata Beutenmuller (nec Walker) Bull. Am. Mus. N. Hist. IX, 210 (1897) (partim).
Olene leucophaea Beutenmuller (nec Abbot & Sm.) Bull. Am. Mus. N. Hist. X, 385 (1898); Holland Moth Book, p. 308 (1903) (partim).
Olene plagiata Dyar (nec Walker) Bull. 52, U. S. Nat. Mus. 260 (1902) (partim); id. Proc. Wash. Ent. Soc. XIII, 17 (partim).

"♂ Head and prothorax darker than the rest of the thorax. *Base of the primaries within the basal line yellow. Costa above this yellow spot darker*

than the rest of the wing, which is cinereous without any green olive scales. Basal line straight between the median and internal nervure. *The outer line approaches the inner on the internal margin.* A large orbicular discal circle. Beneath lighter with an obscure common broad diffused line and a discoidal dot on each wing, much larger on the primaries. Length of body .70; exp. wings 1.42 inch. 'Nonantum' (Harris). Coll. Harris)" Pack.
Clintonii.

"Female. Greyish olivaceous, sparsely sprinkled with black, isolated scales. A dark basal line. Space between this line and the inner median transverse line (transverse anterior) pale brown, dotted sparsely with black scales, cinereous-olivaceous along costa. Transverse anterior line slightly oblique; faint, lost in the brown scales of the sub-basal space which precede it, but distinctly indicated, margined outwardly with pale scales, angulated below costa, below m. nervure twice very distinctly and deeply excavate. Median space cinerous-olivaceous, sparsely sprinkled with black scales, paler along costa and especially on the disc and beyond to the t. p. line where the scales are almost white. On these white scales a large reniform discal spot is indicated by a black rather indistinct encircling line and a central brownish streak. Transverse posterior line (outer median) black, broad, very distinct, nearly straight, very slightly undulate between the nervules, the most prominent projection occurring at the second m. nervule outwardly, and another, rounded inwardly, at above internal nervure. Outside the t. p. line is a wide dark brown shade band irregularly margined outwardly with whitish scales and narrowing greatly below the third m. nervule. *On the interspaces,* superiorly, are *darker brown, longitudinal dashes;* above the internal nervure is a dark brown maculate shade, broadly margined outwardly with white scales. Terminally the wing is cinereous-olivaceous as on median space. A terminal narrow black line, margined inwardly narrowly with whitish scales. Fringes dark, interrupted with pale brown at the extremity of the nervules. Secondaries evenly pale cinereous-brown. A darker shaded discal spot and subterminal band; fringes concolorous with the wing.

"Under surface of both wings similar, but a little paler than secondaries above, very sparsely sprinkled with dark scales; a dark discal spot and subterminal line on both pair; on the primaries the discal spot extends diffusedly upwards along the costa where the subterminal band is most plainly marked.

"Head and thorax olivaceous-cinereous; abdomen darker than secondaries with metallic sub-tufts, as usual in the genus, on the second and third segments above.

"Beneath, cinereous; legs clothed with long grey scales; tarsi spotted outwardly with darker scales.

"Male. Much smaller than the female but resembling it in ornamentation, as is usual in this genus; the brown color on the upper surface of the primaries is less conspicuous, the olivaceous shades are brighter and all the markings are less apparent. The secondaries are darker.

"Exp. male 1.40, female 1.80 inch. Length of body, male 0.60, female 0.90 inch.

"Type Loc. Rhode Island. (Seekonk.) Coll. Mrs. S. W. Bridgham."

G. & R.

LARvA.

"(*before last moult*). Head shining black. Body pale yellowish, variegated with black; a black dorsal line, interrupted on the summits of the posterior segments. Long silky white hairs, with a few black ones arise from the subventral warts. The lateral row (row III) furnishes shorter bristly yellowish hairs; but on joints 2 and 13 gives a long pencil of black hairs. A few black hairs also overhang the head and extend from joint 13. From the subdorsal warts on joints 2-4, 8-11 and 13 arise tufts of plumed white hairs appearing 'mouldy' on the ends, intermixed with bristly yellow hairs. On joints 5, 6, 7, and 12 the warts of rows I and II bear a series of large square black tufts, mixed with white plumed hairs especially at the sides of the tufts, where also a few bristly yellow hairs occur. The tuft on joint is much less black than the others. Dorsally on joints 10 and 11 a median whitish retractile tubercle with flattened top.

"*Last stage*. Head black, whitish above the mouth. Body pale whitish with a yellowish tinge, shaded, marked diffusely with black; a dorsal and a stigmatal band indicated. Two long black pencils of hairs on joints 2 and 13 as in the previous stage. Lateral hairs long, dirty whitish mixed with a few black ones. Dorsal tufts as before except that those on joints 5-8 are now large, square, brown ones, mixed at the sides with white plumed hairs; the tuft on joint 12 still remaining black as previously and contrasting with the others.

"*Cocoon* composed of hair and silk.

"Mature larva on Hickory (Carya) at Rhinebeck, N. Y., June 6, 1887, and young ones on oak (Quercus) Aug. 9, 1887." Dyar.

Fresh bred specimens, especially in the ♂ sex, show a strong olive-green suffusion. In the ♂'s the yellow-brown basal patch is apparently quite variable; in a series of 6 ♂'s before us from New Brighton, Pa., 2 ♂'s are typical, a third has the yellow suffused over the whole basal area, and the remainder show the merest traces of this color. (Pl. I, Figs. 3, 5). A ♂ in the Strecker Coll. labelled 'Md.' is similar, with the addition of a black submedian dash. These specimens present quite a distinct appearance from the typical form, being rougher in squammtion, approaching the pine-feeding forms in this respect. Collectors in these localities would do well to study the early stages as the possibility of a distinct species is by no means improbable.

The t. a. line in ♂ is very irregular and jagged with usually a prominent outcurve just above inner margin; at this point the two lines are often extremely close together, but their relative position is slightly variable. The subterminal brown band is not so prominent as in the

♀, but shows distinct traces of the black dashes, especially between veins 6 and 7, where a dark point juts out and touches the outer margin; a whitish spot at anal angle is present in all specimens before us. The discal spots and subterminal lines of underside are usually distinct.

In the ♀'s the lines are relatively farther apart than in the ♂, the basal patch is less prominent owing to its greater suffusion and the median space is rather paler. The distinctive feature is the dark brown subterminal band with prominent darker transverse dashes. The secondaries vary in depth of color and may be prominently banded with smoky brown or remain almost immaculate. A single ♀ from New Brighton, Pa. (Pl. I, Fig. 6) is deeper in color and has a basal black dash along submedian fold; a similar ♀ from 'Md.' is in the Field Museum (Strecker Coll.).

HABITAT. Boston, Mass. (Packard); Rhode Is. (Grote & Rob.); Yaphank, L. I. (Doll, ex l.) (July 23, 24); N. Jersey (Hy. Edw. Coll.); New Brighton, Pa. (Merrick) (July 26-Aug. 4), Md. (Strecker Coll.).

(b) O. BASIFLAVA MERIDIONALIS subsp. nov. (Pl. II, Figs. 3, 4).

Oryja leucophaea Hubner (nec Abbot & Sm.) Samml. Exot. Schmett. II, Pl. 179 (1820).

Dasychira leucophaea Walker, (nec A. & S.) Cat. Brit. Mus. IV, 870 (1855).

Olene leucophaea Holland (nec Abbot & Smith) Moth Book, Pl. 37, figs. 7 and 8 (1903); Psyche, Pl. X, fig. 3 (1904).

Olene plagiata Dyar (nec Walker) Proc. Ent. Soc. Wash. XIII, 17 (1911) (partim).

Olene achatina Barnes & McDunnough (nec Abbot & Smith) Psyche XVIII, 157 (1911) (larva).

♂ Slightly smaller than in the type form; basal yellow patch not so prominent, more diffused with brown; t. a. and t. p. lines much less irregular and jagged, angle of t. a. line at inner margin not so prominent, often entirely lacking; t. a. line often preceded by a parallel white line about 1 mm. nearer base of wing.

♀ Subterminal band very deep brown, very sharply defined outwardly, contrasting greatly with the pale olivaceous median and terminal areas; t. a. line almost straight, the prominent indentations of the type form almost entirely lacking; basal space to t. a. line often entirely brown, at times more as in type form with olivaceous costal suffusion. Expanse ♂ 30-35 mm.; ♀ 38-48 mm.

Types ♂ and ♀ in Coll. Barnes; Cotypes in Am. Mus. Nat. Hist. and Doll Coll.

This is the species which has generally been confused with *leucophaea* A. & S. We have a long series before us bred by Seifert from ova obtained in Florida. We have also personally collected and bred the larvae on oak at Lakeland, Florida, and received similar larvae from Southern Pines, N. C. The larva (Pl. V, Fig. 7) (vide Psyche XVIII, 157) apparently agrees with that of *basiflava basiflava* Pack., which we only know from Dyar's description. It is possible that a careful comparison may show certain points of distinction. There are seemingly two generations.

HABITAT. Lakeland, Fla. (June 1-7) (types) (McDunnough, Grossbeck); Island Grove, Fla. (ex. ova July 20-Aug. 3rd) (Oct. 25-30) (Seifert) Hogart Land, Fla.; Talahassee, Fla.; Beaufort, S. C. (May 7) (Bradford); Southern Pines, N. C. (larva) (Manee).

O. KERRVILLEI sp. nov. (Pl. I, Figs. 7-9).

♂ Head and prothorax gray, patagia and remainder of thorax deeper in color; primaries very even olive green, sprinkled slightly with black scales; a black basal half-line; basal space even olive-green with very slight dash of paler color below the cell; t. a. line distinct, deep black-brown, rather evenly waved, especially below cell, general course inclined somewhat outward; a slight purple brown shading precedes the t. a. line, defined towards base of wing by a faint whitish line parallel to t. a. line; median space of same color as rest of wing except reniform, which is defined by black on a white patch, this patch not extending to costa; t. p. line distinct, deep black-brown, slightly convex from costa to anal vein where it bends outward to reach inner margin, slightly waved with a small outward projection on vein 4; subterminal and terminal areas not well defined, subterminal line being an obscure irregular whitish line preceded by some slight brown scaling and ending in a distinct white spot above anal angle, between veins 6 and 7 a black dash extends to outer margin, otherwise both areas even olive-green; terminal line black, more or less broken and irregular; fringes slightly checkered. Secondaries deep smoky, paler along outer margin with a darker discal spot and curved postmedian line, fringes unicolorous or slightly checkered. Beneath, pale ochreous, sprinkled with black, primaries with prominent black reniform preceded by dark diffuse shade in cell and with an incomplete subterminal line; secondaries with large dark discal spot and complete subterminal line.

♀ Head, thorax, and abdomen, gray; primaries ground color olivaceous gray, sprinkled with sparse black scales, basal space to t. a. line brown, paler at times along costa; t. a. line outwardly oblique from costa to middle of cell, slightly waved, then straight with prominent inward angle below median vein and outward angle in submedian fold, becoming rather obsolete towards inner margin; the t. a. line may or may not be preceded by a parallel whitish line crossing the brown basal area; median area largely pale, whitish in costal portion with traces of a large bent reniform mark outlined in brown; t. p. line distinct, deep brown, slightly inwardly oblique, straight from costa to vein 5, slightly dentate at veins 3 and 4, waved in submedian fold and bent outward to inner margin; subterminal space brown, the outer margin being sharply defined and very jagged, outwardly oblique from costa to space between veins 6 and 7, where it sends a very prominent dark dash nearly to outer margin, thence incurved with a second less prominent projection between veins 3 and 4, then strongly incurved to near t. p. line, becoming indistinct towards inner margin; this sub-terminal space contains two short black dashes below costa and a broad promi-nent suffused dash between veins 5 and 6 extending inwards almost to t. p. line; both outward projections are suffused with black and there is a further black spot in submedian fold bordered outwardly with whitish; from this white spot a black longitudinal bar may or may not extend backward to base of wing; terminal space olivaceous gray; a faint irregular dark terminal line bordered inwardly partially with white; fringes brown, slightly checkered. Secondaries pale to dark brown with traces of discal spot and subterminal band. Beneath, light brown with faint discal spots and incomplete median lines. Expanse ♂ 37 mm., ♀ 48 mm.

Habitat. Kerrville, Tex. (May, June, Oct.) (Lacey) 2♂'s, 3♀'s. Types, Coll. Barnes; Cotypes, Coll. Lacey.

The ♀'s are very close to *basiflava* Pack. and we should have hesitated to describe the species as new if it had not been for the very distinct appearance of the ♂ sex. Two of the ♀'s before us possess a black dash in submedian fold; in general the ♀'s may be separated from allied forms by the inward costal bend of t. a. line, the extremely marked dark dash-like suffusion in brown subterminal area and the very prominent dash extending to outer margin between veins 6 and 7, the t. p. line is also straighter in general course. The even olive-green appearance of the ♂'s at once separates them from *basiflava*, and allies them with *leucophaea* A. & S. They most closely approach Hubner's figure of *tephra* of any species we have seen but among other things possess a distinct white anal spot on primaries and lack the checkered fringes of secondaries; the ♀ on the other hand is entirely different to Hubner's figure, which shows none of the dark subterminal shading so marked in our species. The fact that a single ♀ is dated "October" would point to at least a partial double brood. Through the kindness of Mr. Lacey the types remain in Coll. Barnes.

O. VAGANS sp. nov.

?Parorgyia clintonii Coquillett (nec. G. & R.) Can. Ent. XII, 45 (1880) (larva).
Olene plagiata Neumoegen & Dyar (nec. Walker) Jour. N. Y. Ent. Soc. II, 58 (1894) Beutenmuller, Bull. Am. Mus. Nat. Hist. X, 385 (1898); Dyar, Bull. 52 U. S. Nat. Mus. 260 (1902); id, Proc. U. S. Nat. Mus. 24, 887 (1904); Dod. Can. Ent. XXXVIII, 53 (1906); Dyar, Proc. Ent. Soc. Wash. XIII, 17 (1911) (partim.).

This is the species which has generally passed under the name of *plagiata* Wlk. (*clintonii* G. & R.) for the past ten years. As however, *plagiata* Wlk. (1855) has been shown to be synonymous with *leucostigma* A. & S. and *clintonii* with *basiflava* Pack., this species is left without a name.

At the present time it is the most puzzling species of the genus owing to the facts that its range is apparently very extended, (it is found in the N. E. states and southward into New York, spreading westward through Southern Canada to the Pacific Coast and southward along the Rocky Mts. into Utah, Colorado and New Mexico), that it can be differentiated into several apparently local races, and that the early stages are practically unknown, leaving it a matter of some doubt whether we are dealing with geographical forms or distinct species.

Owning to the kindness of Prof. T. N. Willing of Saskatchewan University, we have received a blown larva and two ♂ adults which can be definitely associated; a blown larva received from Mr. A. Gibson of Ottawa, without however any definite locality or associations, agrees with our Saskatchewan species in all but a few minor details and leads us to the supposition that this may be the larva of the Eastern form. The larva described by Coquillett under the name of *clintonii* G. & R. and referred by Beutenmuller to *plagiata* Wlk. is seemingly very close to this larva from Ottawa; Mr. J. Doll of the Brooklyn Inst. Museum who saw this larva while on a visit here informs us that he collected a similar larva on Long Island in 1911; the resulting ♀ we have before us and it certainly belongs to this species.

For the above reasons it appears advisable to us to treat all these forms as geographical races of one variable species and leave it to collectors in the different localities to verify or disprove the truth of our statements; we characterize below three subspecies.

(a) O. VAGANS VAGANS subsp. nov. (Pl. III, Figs. 1, 2, and 4).

♂ Head and collar light gray; thorax darker gray-brown; primaries in general appearance gray-brown with a heavy sprinkling of black scales and a distinct greenish tinge (faded specimens lack this); an indistinct dark basal half-line; basal space smoky brown; t. a. line black-brown, upright, somewhat waved but without prominent angles, preceded by a narrow band of olivaceous; median area paler than base of wing, olivaceous, whitish in costal portion with a large reniform outlined indistinctly in dark brown on this white patch; t. p. line rather evenly sinuate, slightly waved, outcurved around cell, bent inwards in submedian fold with an inward angle on vein 1; subterminal space of same color as base of wing, slightly suffused with olivaceous, very irregular on outer margin, broad at costa with traces of two diffuse dark dashes, touching margin of wing between veins 6 and 7 with a dark arrow-like dash, outwardly rounded between veins 3 and 4 then greatly narrowed to inner margin, bordered outwardly with whitish which tends to form a small patch above anal angle; terminal space beyond white shading dark with indistinct terminal broken black line; fringes dark, checkered. Secondaries deep smoky brown with traces of darker discal dot and subterminal banding. Beneath gray-brown with a large lunate discal spot on primaries and a small dot on secondaries and a more or less distinct subterminal line across both wings.

♀ Primaries olivaceous gray, suffused in appearance, rendering the maculation more or less indistinct; basal space shaded with brown especially above inner margin; t. a. line very indistinct with inward angle below cell; median space shaded with white around reniform, otherwise rather evenly olivaceous; t. p. line dark, distinct, inwardly oblique, projected outwardly on vein 4 and at inner margin; subterminal brown area suffused with olivaceous and but poorly defined from the terminal area by an indistinct waved whitish shade which forms a slight white spot above anal angle; an irregular dark terminal line. Secondaries pale brown with traces of discal dot and subterminal line. Beneath light brown with markings as in ♂. Expanse ♂ 38 mm., ♀ 48 mm.

HABITAT. ♂, St. Johns, Que. (July 1 and 11) (Chagnon) (type and cotype); ♀ Windsor Mills, Que. (ex larva on beach, July 4) (Winn); Yaphang, L. I. (July 22) (Doll) (type and cotype). The types are, through the kindness of Messrs. Chagnon and Winn, in Coll. Barnes; ♂ cotype with Mr. Chagnon; ♀ cotype in Mus. Brook. Inst.

Although we have a good series of what appears to be this species before us from various Eastern localities, we refrain from making more than a single cotype of each sex, so that in the event of an error on our part the name may be easily fixed. Besides the type localities we have specimens from White Mts., N. H. (Coll. Hy. Edw.); Winchendon, Mass. (Russell), Franconia, N. H. (Coll. Doll); Maine (Coll. Angus); Big Indian Valley, Catskills, N. Y. (Pearsall); Geneva, N. J. (Meyer); Montreal, Que. (Winn).

The ♂'s, especially when faded, very closely approach the form of *basiflava* without yellow at base of wing; in a general way they are however more even and smoother in coloration and the space between t. a. and t. p. lines on inner margin is wider; the former line is also much less jagged. The ♀'s may be distinguished from *basiflava* ♀'s by their suffused appearance and lack of definite markings in basal and subterminal areas, single specimens, especially when aged, are however rather hard to place and breeding will be necessary to determine the range of variation of the two species.

We append a description of what we imagine to be the larva of this form, drawn up from the specimen already mentioned as coming from Mr. Gibson; we also figure it (Pl. V, Fig 6).

"Head black; body marbled laterally with white; dorsal tufts on segments 4-7 squarely cut, brown, intermingled with white hairs; tuft on segment II black, narrow, considerably raised above two white bunches of plumed hair supporting it on each side; two moderately long black lateral hair pencils in front and similar ones behind; other warts with small tufts of plumed white hair, those of dorsal and subdorsal row intermixed with ochreous bristles; on segment 4 the tubercle of the supra-spiracular row (III) bears a single black central plumed hair in addition to the others; the subspiracular row (IV) has 2 or 3 plumed black central hairs to each tubercle with segment 4 bearing 5 or 6 such hairs; spiracles pale ochreous, black rimmed; eversible glands coral red."

(b) O. VAGANS GRISEA subsp. nov. (Pl. III, Figs. 5, 6).

Olene plagiata Barnes & McDunnough (nec Walker) Cont. N. Hist. N. Am. Lep. Vol. I, No. 4, Pl. I, figs. 1 and 2 (1912) (types).

Much grayer and without the noticeable greenish tinge characteristic of the type form; slight traces of olivaceous only in median space; reniform situated on a more or less distinct white patch; gray shade preceding t. a. line often distinctly marked in ♂; line of demarcation between subterminal and terminal areas only distinct at costa as a whitish shade line; white supraanal spot very indistinct or missing; terminal broken black line edged inwardly with white. Secondaries of ♂ paler than in type form.

HABITAT. Eureka, Utah (26-31 July) (Spalding) (10 ♂'s); Provo, Utah (July 9) (Spalding) 1 ♀. Types, Coll. Barnes.

A series of 9 ♂'s and 1 ♀ before us from Glenwood Spgs., Colo., taken in July and August are slightly darker in coloration, with almost no white; a ♂ from Beaver City, Utah, (Pl. III, Fig. 7) received through Mr. Doll agrees with these Colorado specimens and through the kindness of the same gentleman we have had the chance of examining two immature larvae (½ grown) from the same locality; these differ from the Eastern and Saskatchewan forms in having the lateral body-clothing white, entirely without black plumed hairs; the dorsal tuft on segment 11 is very black and long, giving almost the appearance of a strong hair pencil; the other dorsal tufts mouse-gray and squarely cut, the anterior and posterior hair pencils similar to those of the other subspecies. Not knowing the larvae of the Eureka, Utah form we can only refer this larva provisionally to *grisea;* they were, according to Mr. Doll, plentiful on oak. We have very similar ♂'s also from Cartwright, Man. (Heath), (Pl. III, Fig. 3) Winnipeg,

Man.; Pincher, Alta (Willing); Kaslo, B. C. (July 8-20) (Cockle); Oak Creek Canyon, Colo. (Doll). As before however, we have restricted the types to a single locality to avoid confusion.

Finally we have before us 1 ♂ and 3 ♀ from the old Bolter Collection from New Mexico, and a ♀ in Coll. Barnes from Durango, Colo., which are deep brown in tint and apparently represent a good race; we await fresher material before venturing to bestow a name.

(c) O. VAGANS WILLINGI subsp. nov. (Pl. III, Fig. 8).

♂ Much deeper in coloration than the two preceding forms, of a general suffused blue-black appearance; t. a. line broad, waved, deep purple-brown; basal space dark with slight metallic green tinge; t. p. line similar in color to t. a. line; median space paler, especially around reniform, with or without greenish suffusion in lower portion; subterminal line vague, whitish, not so prominently angled as in type form; distinction in coloration between subterminal and terminal areas not marked; terminal line black, irregular, with slight inward white shading. Secondaries even smoky black with vague discal dot. Expanse 35 mm.

HABITAT. Humboldt, Sask. (July 13 and 27) (Willing) 2♂. Type, Coll. Barnes; Cotype, Coll. Willing.

We take much pleasure in naming this form after Prof. T. N. Willing who has kindly sent us specimens for study. We have 6 very similar ♂'s from Hymers, Ont. (Dawson) (Pl. III, Fig. 9) and a single ♂ from Winnipeg, Man. (Coll. Winn). We have chosen the Saskatchewan specimens as types for the reason that we have a larva (Pl. V, Fig. 5) of this form before us from the same region which we describe as follows:

Head black; body marbled laterally with whitish; dorsal tufts on segments 4-7 mouse gray with a slight admixture of white laterally; tuft on segment II black, edged by the white plumed hairs proceeding from tubercle III; black anterior and posterior lateral hair pencils; remainder of body clothing consisting of plumed whitish hairs admixed with white bristles from tubercles II and III; tubercle IV (subspiracular) bears a single black central plumed hair (on segment IV, 2 black hairs); spiracles ochreous rimmed with black; eversible glands coral red." Length (blown) 38 mm.

This larva only differs from that described under the type form in the more even mouse-gray color of tufts and the fact that the lateral tubercles bear fewer black plumed hairs; the number is apparently constant, as several minature larvae from the same source agree in this respect with the full grown specimen. The food plant is poplar.

O. ATRIVENOSA Palm. (Pl. VII, Figs. 5, 6).

Parorgyia atrivenosa Palm. Jour. N. Y. Ent. Soc. I, 21, Pl. I, fig. 5 (1893).
Olene leucophaea var. *atrivenosa* Neumoegen & Dyar, Jour. N. Y. Ent. Soc. II,
  58, (1894); Beutenmuller, Jour. N. Y. Ent. Soc. II, 58, footnote (1894),
  (good species); Dyar, Bull. 52, U. S. Nat. Mus. 260 (1902).
Olene atrivenosa Dyar, Proc. Ent. Soc. Wash. XIII, 18 (1911).

"Male. Primaries grayish fuscous, with the veins marked with blackish-brown scales. The inner half of the wing is somewhat paler, with several dirty, white, irregular patches. On the apical third is a narrow, curved, blackish transverse band, which forms an angle before it reaches the inner margin. Before the outer margin is an irregular, grayish, patch-like band. Hind wing grayish fuscous, with an obsolete discal spot of a deeper color. Before the outer margin is a distinct broad band of a lighter color, running from a little below the apex, nearly to the anal angle.

Underside, dirty, grayish white, with a broad transverse smoky-gray band across the wings. Also a discal spot of the same color. Thorax and body, grayish fuscous. Expanse 25 mm.

Female. Differs from the male by having the transverse band on the primaries almost obliterated. Band before the cilia on the secondaries indistinct. Otherwise same as male. Expanse 33 mm.

One male and one female. Hab. Red River region, Arkansas. Types Coll. Chas. Palm."

This species is entirely unknown to us. As far as we can judge from the photographs received from Mr. C. Palm it would seem best associated with the *basiflava* group.

O. LEUCOPHAEA A. & S. (Pl. II, Figs. 1, 2).

Phalaena leucophaea Abbot & Smith, Lep. Ins. Ga. II, 155, Pl. 78 (1797).
Dasychira leucophaea Walker, Cat. Lep. Het. Brit. Mus. VII, 1738·(1856).
Parorgyia leucophaea Packard, Proc. Ent. Soc. Phil., III, 333 (1864); Grote,
  Can. Ent. III, 124 (1871); id. C. Ent. XIX, 113 (1887) (?).
Olene leucophaea Dyar, Bull. 52, U. S. Nat. Mus. 260 (1902); Barnes & Mc-
  Dunnough, Psyche XVIII, 158, Pl. XIII (1911).

Until we bred this species in 1911 from larvae so closely approaching Abbot's figure in color as to leave no doubt of their identity, this species had remained unrecognized or wrongly identified. As the name is based solely on a figure we append the following description drawn up from 1 ♂ and 2 ♀ ♀ in Coll. Barnes. They are the only specimens known to us.

♂ Head and thorax gray, patagia and thoracic tuft darker, latter tipped with orange. Primaries, ground color rather even purplish gray; a black basal half-line; basal space to t. a. line largely brown with the exception of a broad purple-gray costal patch, extending from base of wing to a waved whitish line preceding and parallel to the t. a. line, and some purplish scaling along inner margin; below the cell faint traces of a black longitudinal dash ending on the white line in a small round patch, below this patch the white line is indistinct; t. a. line almost perpendicular, with three outward projections, in the cell, in the submedian fold, and just above the inner margin; median space rather even purple-gray, paler at costa; a large bent reniform partially outlined in black; t. p. line bent inwards at costa with a slight outward projection between veins 3 and 5, then incurved and sharply bent outwards just above inner margin; subterminal space shaded with brown and with traces of suffused dark longitudinal streaks most prominent near costa; the outer margin of subterminal space clearly defined but not very jagged, broadest at costa with short outward ray between veins 6 and 7, thence gradually narrowing to vein 2 from which it is parallel to t. p. line; terminal space purple-gray tinged with whitish near apex and with prominent white spot above anal angle; a terminal dark line broken and irregular towards anal angle. Secondaries dark smoky brown with traces of a large discal dot. Beneath smoky brown with the usual discal dots and postmedian lines.

♀ Much paler than ♂, olivaceous gray presenting a generally soft and suffused appearance; the basal brown area is also paler, defined as in the ♂; t. a. line with more prominent angle in the cell, the white line preceding it more distinct and continued to inner margin; costal portion of median area white, contrasting with the remaining portion which is olivaceous gray; reniform almost indistinguishable; t. p. line waved, inwardly oblique from costa to vein 5, then bent outward to vein 3 and gradually curving inward to anal vein, attaining inner margin by a sharp curve outward; subterminal space brown with numerous suffused dark dashes, broad from costa to vein 3, projected somewhat outwardly between veins 6 and 7 and 3 and 4; terminal area as in ♂ but paler, with distinct white anal dot, preceded in subterminal area by dark shading. Secondaries pale smoky with suffused subterminal dark banding. Beneath pale gray with traces of discal spots and postmedian lines. Expanse ♂ 38 mm.; ♀ 43 mm.

LARVA.

"Dull yellow-ochre. Head black with pale yellow mouth parts; body gray-green, marbled slightly with ochreous and with a broad blackish dorsal band on abdominal segments V-IX; dorsal tufts on abdominal segments I-IV and VIII dark ochreous, the latter with a long black hair pencil; two anterior and two posterior lateral black hair pencils; tubercles pale ochreous, with numerous plumed hairs, similar in color to the tufts; eversible conical glands of abdominal segments VI and VII pale yellowish. Prolegs flesh color with dark lateral plate. Spiracles pale cream with black vein. Length 40-55 mm. The ♀ larva undergoes an extra skin-shedding." B. & McD.

HABITAT. Georgia (Abbot & Smith); Southern Pines N. C. (Manee).

O. ATOMARIA Wlk. (Pl. IV, Figs. 5, 6; Pl. VI, Figs. 1, 2; Pl. VII, Fig. 2).

Dasychira atomaria Walker, Cat. Lep. Het. Brit. Mus., VII, 1739 (1856) (♀ nec ♂).

Parorgyia achatina Packard (nec A. & S.) Proc. Ent. Soc. Phil., III 333 (1864).

Parorgyia obliquata Grote & Robinson, Proc. Ent. Soc. Phil. VI, 4 Pl. 1, fig. 4. (1856).

Parorgyia parallela Seifert (nec G. & R.) Ent. Amer. III, 96 (1887) (partim); Packard (nec G. & R.) 5th Rep. U. S. Ent. Comm. 135-6, figs. 42, 43, Pl. 35, fig. 3 (1890) (larva).

Parorgyia achatina var. *obliquata* Dyar, Psyche VII, 136 (1894) (larva).

Olene achatina var. *tephra* Neumoegen & Dyar (nec Hubner) Jour. N. Y. Ent. Soc. II, 57 (1894).

Dasychira plagiata Beutenmuller (nec Walker) Bull. Am. Mus. N. Hist. IX, 210 (1897) (partim).

Olene achatina Beutenmuller (nec A. & S.) Bull. Am. Mus. N. Hist. X, 384 (1898).

Olene achatina var. *tephra* Dyar (nec Hbn.) Bull. 52, U. S. Nat. Mus., 260 (1902).

Olene obliquata Dyar, Proc. Ent. Soc. Wash., XIII, 16 (1911).

With a colored figure and a photograph (Pl. VII, Fig. 2) of the type ♀ of *atomaria* Wlk. as well as a specimen compared for us by Sir. Geo. Hamson with type and a photograph of the two type ♀'s of *obliquata* G. & R. (Pl. VI, Figs. 1, 2) before us, we have no hesitation in adopting the above synonymy. We also have received photographs of the ♂ and ♀ in the Harris Collection referred to by Packard as *achatina* (Proc. Ent. Soc. Phil. III, 333); the ♂ is the form *parallela* G. & R. and the ♀ is typical *atomaria* Wlk.

The species occurs in two forms, with and without a black basal dash, as has been conclusively proved by the breeding experiments of Seifert. The form with black basal dash has been described as a good species by Grote and Robinson under the name *parallela;* it is sufficiently distinct to warrant a retention of the name. The ♂'s have been briefly described by Seifert, and figured by Packard under the name *parallela*. The early stages are described by Packard (his final stage is evidently erroneous) from ova from Newburgh, N. Y.; Dyar also gives notes on some of the larval stages and Seifert has worked out the life history in detail (Ent. Amer. III, 93) from larvae found in Green Co., Catskills, N. Y.

Fresh specimens show a distinct olivaceous shading on primaries which is rather liable to fade leaving the brown hues more prominent, especially in the basal area. The characteristic features of the species are found in the subterminal brown area which is less jagged on its outer margin than in *basiflava* Pack. and also more suffused and defined outwardly more or less distinctly with whitish; the prominent dark dashes and the white supra-anal spot are wanting. The t. a. line in ♀ has usually a very prominent outward angle in the cell. We append Grote & Robinson's description of the ♀ as being more satisfactory than Walker's, and have added a description of the ♂ by ourselves.

"♀. Evenly olivaceous-cinereous, very sparsely irrorate with black scales. Transverse lines pale brown. Inner median line irregularly dentate, brown. A brown discoidal streak around which the scales are faintly and irregularly paler than elsewhere. Outer median line nearly straight and even, brown, shaped much as in *P. Clintonii,* but with still less prominent inflections and succeeded by a similarly colored rather paler brown shade band, which is equally wide but rather more diffuse superiorly, where it is faintly margined with paler scales. A terminal brown line further from the margin and more irregular than in *P. Clintonii.*

Secondaries pale grayish-brown, darker than in *P. Clintonii;* a diffuse darker subterminal shade band and very faint discoloration.

Under surface slightly darker than upper surface of secondaries. On anterior wings a faint, discal, darker discoloration and a discontinued, oblique, even, subterminal, shade band. On the secondaries a discal undefined spot and a distinct, even, oblique dark band crossing the wing from within the apex to anal angle. This band runs within and at variance with its analogue on the upper surface, as can be seen by holding the specimen to the light; its obliquity and peculiarity has suggested the specific name. On the primaries, also, the subterminal band is not produced immediately beneath the outer median line.

Head, thorax and appendages covered with mixed grey scales. Abdomen paler than secondaries. Exp. ♀ 2.00 inches. Length of body, 0.85 inch.

Habitat. Rhode Island. (Seekonk). Coll. Mrs. S. W. Bridgham.

The more simple ornamentation of this species will quickly distinguish it from *Parorgyia Clintonii* nob. The male is unknown to us." G. & R.

♂. Smaller and darker than ♀, strongly tinged with olivaceous in fresh specimens; a dark basal half-line; basal space to t. a. line largely brown, tinged with olivaceous along costa and inner margin; t. a. line irregular, dentate, with three outward projections,—in the cell, in the submedian fold and above the inner margin—preceded by a rather indistinct white parallel line; median space shaded with white around reniform, which is fairly distinctly outlined in deep brown; t. p. line slightly waved, rather evenly outcurved from costa to below cell, then incurved in submedian fold and bent outward at inner margin; subterminal space suffused brown, shaded with olivaceous, bordered by an indistinct, irregularly dentate whitish s. t. line; terminal space evenly olivaceous with a dark irregular terminal line; fringes checkered, brown. Secondaries deep smoky brown with lighter fringes and traces of discal spot and postmedian line. Expanse 35 mm.

LARVA.

"Mouse gray, feathered and soft. Body dark slate, almost black, velvety, stigmatal region light yellowish gray. Dorsal tufts on 4th, 5th, 6th, 7th and 11th segments dark, those on 4th and 11th being almost black. A pair of long black lateral hair-pencils on 1st segment and a few single black hairs from anal segment. Suprastigmatal warts on segment 1 without feathery hair; all other warts with bushy, rounded tufts of feathery hair. Eversible glands on 9th and 10th segments amber-colored. Spiracles whitish; legs whitish, rather hairy.

PUPA.

Light brown, wing cases and stigmata dark brown; thoracic region, segment joints and cremaster are brown. All the warts, even the pedal line, seem to be retained on the abdominal segments as minutely granulated patches, covered with short hair. The six dorsal warts on 5th, 6th and 7th segments are represented by 6 rosette-shaped, lichen-like formations of yellowish-gray color, the two warts on each segment almost confluent with an oval patch. The chitin cover is very fragile." Seifert.

(a) O. ATOMARIA forma PARALELLA G. & R. (Pl. IV, Figs. 7-9, VI figs. 3, 4).

Parorgyia parallela Grote & Robinson, Proc. Ent. Soc. Phil. VI, 5, Pl. I, fig. 5 (1866); Lintner, Entom. Contr. III, 129 (1874) (larva). Seifert, Ent. Amer. III, 93 (1887) (life history).
Parorgyia achatina Dyar (nec A. & S.) Psyche VII, 136 (1894).
Olene achatina Neumoegen & Dyar (nec A. & S.) Jour. N. Y. Ent. Soc. II, 57 (1894); Beutenmuller, Bull. Am. Mus. N. Hist. X, 384, Pl. XVII, fig. 18 (1898); Dyar, Bull. 52, U. S. Nat. Mus. 260 (1902); Holland, Moth Book, Pl. XXXVIII, fig. 9 (1903).

Grote & Robinson's description of the ♀ is very full and accurate; specimens vary in the amount of brown suffusion in the lower portion of median area; the black shading on the veins may be present or absent. The ♂ is similar to that of *atomaria* Wlk. with the addition of the basal black dash.

"♀. Anterior wings pale olivaceous-cinereous, much clouded with brown and sparsely sprinkled with black scales. Basally the costal half of the wing is olivaceous cinereous; the median nervure is covered narrowly along its length with black scales which are prolonged along the fourth m. nervule. Below, a broader black longitudinal stripe runs from the base of the wing to beyond the t. p. line along the sub-median fold, and is connected with the dark scales along the median nervure by an oblique black line (appearing as if covering a vein) at about its middle. This is part of the inner median line which is dentate and much as in *P. Clintonii* but less distinct. The inner median line is preceded by brown scales which stretch, beneath the median nervure, from the base of the wing across the median space centrally and on both sides of the outer median line, not extending inferiorly below the black longitudinal stripe. The discal space is covered with white scales but the outlines of the spot are indeterminate. The outer median line is distinct, black, strongly marked and, while slightly, excavate and irregular, is without prominent inflections. Terminally and below the prominent longitudinal stripe the wing is covered with pale olivaceous-cinereous scales. Terminal line brown, much as in *P. obliquata* nobis. The fringes are much alike in all three species.

Secondaries pale brownish, no discal spot apparent; a neatly defined, irregular, narrow blackish band, which is further removed from the external margin than in either of the preceding species and more concise. Under surface a little darker than upper surface of secondaries, no perceptible discal spots; a common distinct dark band; on the secondaries it is wider than that on the upper surface, but covering it and entirely analogous to it.

Head and thorax cinereous; abdomen concolorous, or nearly so, with secondaries; beneath, darker, as are the legs, these latter with darker maculations on tibiae and tarsi outwardly. Under thoracic surface griseous.

Exp. female 2.00 inches. Length of body, 0.95 inch.

Type Loc. Rhode Island. (Seekonk.) Coll. Mrs. S. W. Bridgham.

A very distinctly marked species; the dark parallel longitudinal stripes on the upper surface of primaries suggested the specific name. This species is evidently allied to *Parorgyia achatina*, but, judging from Abbot's figure, the Southern species seems sufficiently distinct." G. & R.

HABITAT. Franconia, N. H. (Coll. Doll); Rhode Is. (G. & R.); Dutchess Co., N. Y. (Coll. Doll); New York (Seifert), (Angus); Big Indian Valley, Catskills, N. Y. (June 2) (Pearsall); Sherborn, Mass., (Aug. 15) (Coll. Merrick); Boston, Mass., (Hy. Edw. Coll.); Medford, Mass., (Hy. Edw. Coll.); Geneva, N. J., (Aug. 16) (Meyer); N. J. (Beutenmuller); White Mills, Pa., (Aug. 2) (Coll. Doll); New Brighton, Pa., (Aug. 1-15) (Merrick); Lititz, Pa., (Aug. 16) (Coll. Doll); Nebraska (Coll. Hy. Edw.) (?) Tallahassee, Fla., (Koebele) (Coll. Doll) (?).

The last two localities require confirmation.

O. CINNAMOMEA G. & R. (Pl. IV, Figs. 1-4).

Parorgyia cinnamomea Grote & Robinson, Proc. Ent. Soc. Phil., VI, 6 Pl. I, fig. 6 (1866); Dyar, Psyche VII, 137 (1894) (good sp.).

Olene cinnamomea Neumoegen & Dyar, Jour. N. Y. Ent. Soc. II, 57 (1894); Dyar, Proc. Wash. Ent. Soc. XIII, 17 (1911).

Olene achatina var. *cinnamomea* Dyar, Bull. 52 U. S. Nat. Mus. 260 (1902).

"♀. Brown. Basally the anterior wings are entirely dull brown. Median lines brown, shaded narrowly on either side, with pale scales. Median space superiorly covered with pale whitish scales, on which a brown reniform ringlet obscuredly indicates the discal spot. Inferiorly, above internal margin, a bluish black or cinereous scale patch. Below the discal pale patch the median space is brown, paler, but nearly concolorous with the rest of the wing. The outer median brown line is projected outwardly slightly at second m. nervule and roundedly so, inwardly at above internal nervure. Beyond the outer median line the wing is entirely brown to external margin, but the terminal half is scattered over sparsely by cinereous scales, especially at apex and internal angle; fringes dark.

Secondaries concolorous, umber brown, paler at base. Very faint traces of band and discoidal spot. Under surface more brownish than secondaries above. On anterior pair a discal spot and subterminal band are faintly indicated, the latter apparent at costa. Secondaries with hardly a trace of either, except that there appears a faint discoloration along the discal cross-vein.

Head and thorax dark umber brown; under thoracic surface and legs clothed with cinereous scales.

Exp. female 1.40-1.80 inch. Length of body, 0.75 inch.

Type Loc. Rhode Island. (Seekonk) Coll. Mrs. S. W. Bridgham.

A smaller species than its congeners and easily known by its rich brown, nearly concolorous primaries." G. & R.

Whether this is a good species or variety of *atomaria* cannot be determined at the present moment as nothing is known of the larva. Mr. Winn writes us that the species was common at Biddeford, Me., but that he always arrived too late in the season to find any larvae. From the material before us (10 ♂'s, 4 ♀'s) the species appears to be distinctly northern in its range, although Dyar records it from Cocoanut Grove, Fla., possibly an error. The ♀'s are very distinct in appearance with their coppery brown coloration and paler median space; one ♀ before us from Biddeford, Me., (Pl. IV, Fig. 2) is entirely suffused with brown leaving but a trace of the maculation visible; the ♂'s agree with *atomaria* ♂'s in general markings but are considerably browner.

HABITAT. Bangor, Me., (July 16) (Coll. Doll); Biddeford, Me., (July 10-20) (Winn); Sherborn, Mass., (Aug. 7, 26,) (Coll. Merrick); Winchendon, Mass., (Aug. 10) (Russell); Dutchess Co., N. Y., (July 20) (Coll. Doll); Minn. (Hy. Edw. Coll.).

O. MANTO Stkr. (Pl. V, Figs. 1, 2; Pl. VII, Fig. 4).

Parorygia manto Strecker, Lep. Rhop. Het. Supp. 3, p. 29 (1900).

We recognize three races of this species as suggested by Dyar (Proc. Wash. Ent. Soc. XIII, 19); we do not know the var. *interposita* Dyar and more material will be necessary to determine its status; the species we figured (Psyche XVIII, 157) under this name we are inclined, after seeing Strecker's type, to place under *manto* Stkr.

(a) O. MANTO MANTO Stkr.

Parorgyia manto Strecker, Lep. Rhop. Het. Supp. 3, p. 29 (1900).
Olene leucophaea var. *manto* Dyar, Bull. 52, U. S. Nat. Mus. p. 260 (1902).
Olene manto Dyar, Proc. Wash. Ent. Soc. XIII, 18 (1911).
Olene interposita Barnes & McDunnough, (nec Dyar) Psyche XVIII, 158, Pl.
    XIII (1911) (larva).

"Male head and body brownish gray, paler beneath. Primaries rich chestnut brown, a trilobed narrow white subbasal line. An irregular black t. a. line. A black t. p. line curved irregularly inwardly between veins 1 and 4, heaviest from vein 4 to costa. A large kidney-shaped discal spot surrounded by a white line, from this to the costa between the t. a. and t. p. lines white sprinkled with black; on the inner margin between these lines some white scales. A sinuate white submarginal line exterior to which the wing is hoary. Fringe brown, paler at terminations of veins. Secondaries dark brown. Fringe paler. Under surface rather pale smoky brown crossed by a regular diffuse mesial shade. Distinct discal lines on all wings, those of primaries joined exteriorly by a pale spot. Fringes as above.

Expands 1 1/4 inches.

Type Loc. One male, Stewart Co., Georgia, A. W. Latimer.

All the markings on this species are bright and sharply defined." Strecker.

The larva apparently agrees with that of *montana* Beut.; it feeds on various species of pine. The species is probably double-brooded.

HABITAT. South Pines, N. C., (Apr. 24-30, May 8-15) (Manee); Stewart Co., Ga., (Strecker); Hastings, Fla., (May); Fort Meade, Fla., (Feb.) (Dorner).

(b) O. MANTO INTERPOSITA Dyar.

O. interposita Dyar, Proc. Wash. Ent. Soc., XIII, 18 (1911).

"Similar to *manto* Strecker, but more uniformly brown. Fore wing suffused with brown, the lines black, distinct, irregularly crenulate and rather broad. A white cloud in the discal area, defining the brown-filled oblique reniform; subterminal line pale, waved, followed by a grayer terminal area. In the female the median space is largely gray. In *manto* the terminal space is nearly clear gray to the margin and a narrow wavy white line crosses the basal space. The discal region lines and brown ground are very similar in the two forms."
Dyar.

HABITAT. Tryon, N. C., (Fiske).

(c) O. MANTO MONTANA Beut. (Pl. VI, Figs. 8, 9).

Olene montana Beutenmuller, Bull. Am. Mus. N. Hist. XIX, 585 (1903); Dyar,
    Proc. Wash. Ent. Soc. XIII, 19 (1911).

"♂. Fore wings deep smoky brown with a slight violaceous lustre Transverse lines and discal spot almost obscured by the ground color. A lighter brown shade at the middle, from the base to the transverse anterior line. Discal mark elongate, upright, curved inwardly, black, outlined, and with a few white scales. Transverse anterior line vertical, black, with three outward curves. Transverse posterior line, strongly curved outwardly around the discal mark, bending inwardly at about the middle, thence almost straight to the inner margin. A little above the inner angle is a distinct white spot. Line at base of fringes black, with two short teeth inwardly above the angle. Hind wings uniformly smoky brown. Fringes concolorous. Forewings beneath almost uniform smoky brown, paler along inner margin, and with a faint indication of a dark transverse shade. Hind wings beneath smoky brown with a darker discal mark and median shade. Head and thorax smoky brown, the latter slightly lighter brown anteriorly. Abdomen paler than the thorax with two bunches of metallic brown hairs on the anterior segments, and a similar bunch on the disc of the thorax. Antennae smoky brown. Expanse 32 mm.

♀. Fore wings deep sepia brown, basal region at costa and inner margin somewhat smoky brown. Middle part sepia brown with a smoky brown shade. Outer part dark brown. Transverse lines distinct, black, similar to those of the male. Discal mark filled with clear white, and marked with this color on the outer and inner parts. Marginal line black, marked with a little white inwardly, especially near the anal angle. Apex slightly tipped with lighter brown, with an indistinct double-toothed shade beneath it. White spot near anal angle a little larger than in the male. Hind wings uniform smoky brown. Underside wholly smoky brown, with slight indications of a darker median shade on each wing. Head, thorax, and abdomen same as the male. Expanse 36 mm.

LARVA. Head jet black, shining. Body black; brown on the back of the 8th to 12th segments with a short, cylindrical, cherry red protuberance on the 9th and 10th segments. Hairs on the sides of the body brown and black, with a black, plumose hair on each tuft. Bunches of hairs on the 4th to 7th segments inclusive, deep black, with short white, plumose hairs on each bunch at the sides. Hairs on the warts short, white or yellow, plumose. Warts on the 1st to 3rd segments inclusive, brown; on each side brown. On each side of the 1st segment is a long black pencil directed forward and a similar one on the 11th segment directed backward. Underside and legs brown. Length, 40 mm.

Food-plants. Balsam fir (Abies frazeri) and black spruce (Abies mariana).

COCOON. Oblong, oval; composed of silk mixed with the hairs of the larva. Length 18-20 mm., width 8-9 mm.

Type Loc. Summit of the Black Mountains, North Carolina, altitude about 6000 to 6715 feet.

Five immature larvae of this interesting species were collected on June 7. Two of these matured and spun cocoons on June 29, and pupated June 30. On July 8 a female and on July 11 a male emerged. Two larvae were preserved in alcohol, and the remaining one escaped." Beutenmuller.

We only know this form from photographs of the types.

74

O. PINI Dyar. (Pl. II, Figs. 5, 6).

Olene pini Dyar, Proc. Ent. Soc. Wash. XIII, 19 (1911).

"Fore wing gray dusted with black and lightened by white markings, shaded with brown in basal space and between the outer and subterminal lines; a small black line at the base; median lines black, distinct crenulate, the outer angled inwardly on vein 1; discal mark a black reniform, open and more or less broken into two black bars, lightened by white edgings; subterminal line white, waved, with a white spot above tornus; terminal line black, crenulate, somewhat drawn back from the margin; narrow white edgings to both lines. Hind wing brown-gray with discal mark and outer narrow line more or less well defined. Expanse ♂ 30 mm., ♀ 35 mm.

Seven ♂'s and seven ♀'s, North Saugus, Mass., bred from larvae on pine by Mr. W. F. Fiske, Mr. H. M. Russell and myself. Also a ♂ which I take to be the same species labelled 'Cornings Farm, Gray' that is probably from near Albany, N. Y. This specimen has a black submedian bar. Also a ♂ and ♀ labelled 'Sharon, Aug. 1, 1873, July 20, 1874' which are brown and faded looking and without the sharp contrasts of the fresh specimens.

Type No. 13466. U. S. Nat. Mus.

The larva is red-brown or blackish gray, with many plumed white tufts and lateral plumed black hairs; a pair of pencils in front, a pair behind and a single one accompanying the tuft of joint 12; tufts gray, intermixed with plumed white hairs." Dyar.

We have a single ♂ and ♀ before us which apparently belong to this species. They are characterized by the pale median space. The ♂ resembles Dyar's specimen from Cornings Farm in possessing a submedian black bar, which does not however extend across the median pale area. The ♂ is simply labelled "Mass."; the ♀ is from the Catskills Mts., N. Y., (Geo. Franck).

HABITAT. Mass.; Albany, Catskills, N. Y.

O. PLAGIATA Wlk. (Pl. II, Fig. 7; Pl. VII, Fig. 3).

Edema plagiata Walker (nec Walker 1855) Cat. Lep. Het. Brit. Mus. XXXII, 427 (1865); Neumoegen & Dyar, Jour. N. Y. Ent. Soc. II, 173 (1894).
Parorgyia plagiata Grote & Robinson, Tr. Am. Ent. Soc. II, 86 (1868).
Symmerista plagiata Dyar, Bull. 52, U. S. Nat. Mus. p. 252 (1902).
Olene pinicola Dyar, Proc. Ent. Soc. Wash. XIII, 20 (1911).

As stated in the introduction we have received from Sir Geo. Hampson a colored figure and a photograph of Walker's type (Pl. VII, Fig. 3), which is a ♀, not a ♂ as stated. There is no doubt about its being an *Olene*. The figures, combined with Grote & Robinson's remarks on the type, lead us to place it among the pine feeding species and we think Dyar's *pinicola,* if we have correctly identified this species, must fall as a synonym; Dyar's descriptions, however, while noted for their brevity, are not always equally clear. We have before us ♂'s from White Mts., N. H.; Ottawa, Ont.; Hymers, Ont.; and Banff, Alta., which, according to the description, should be *pinicola* and which also greatly resemble the figure of *plagiata*. The type locality of this latter being Orillia, Ont., there is every likelihood that our Ontario specimens are correctly identified. The only other species we know which occurs in this territory is *vagans* B. & McD. and the heavy black lines crossing the wings as well as a slight basal dash in the figure of the type of *plagiata* preclude association with this species.

"♂. Cinereous fawn-color; under side abdomen and hind wings cinereous. Palpi porrect, slender, fringed, not extending beyond the head; third joint extremely small. Antennae slightly pectinated. Thorax with two blackish bands. Legs moderately long; femora and tibiae fringed; hind tibiae with four short spurs. Fore wings minutely black speckled with a black mark at the base and with four broad transverse blackish lines; first and second lines dentate and undulating; second contiguous on its inner side to a white costal patch which is traversed by two brown lines and by an intermediate luteous line; third deeply undulating with a white spot on its outer side near the interior border; fourth marginal. Length of the body 6 lines; of the wings 10 lines. Orilla, West Canada (Bush)."

None of the specimens before us are very perfect and prevent us from amplifying Walker's description. There seems to be considerable variation in the amount of blackish suffusion on primaries. Some of our specimens have the maculation quite obscure in consequence; two specimens from Spruce Brook, Newfoundland, Aug. 10th, from the Coll. Brook Inst. Mus. which we refer here are especially noticeable in this respect. More material may show that these constitute a distinct race.

The larva is probably a pine feeder but is at present undescribed.

HABITAT. Spruce Brook, Nfland., (Aug. 10th); White Mts., N. H., (Coll. Am. Mus.); Ottawa, Ont., (Germain) Hymers, Ont., (July 16-23, Aug. 9) (Dawson); Banff, Alta., (Aug. 2) (Fletcher); Douglas Co., Wis., (a/c Dyar).

O. GRISEFACTA Dyar. (Pl. II, Figs. 8, 9).

Olene grisefacta Dyar, Proc. Ent. Soc. Wash. XIII, 20 (1911).

"Light gray, coarsely dusted with black on a nearly white ground that is irregularly shaded with luteous; lines broad black, the inner coarsely waved, the outer crenulate; discal mark broad, black-outlined; reniform broken above;

subterminal line lost in the general diffusion of the markings or indicated by black inner markings; white spot above tornus distinct, but not contrasted, resembling the white about the discal mark; terminal line black, crenulated, receding from the margin. Hind wings rather light gray, with faint discal spot and outer line. Expanse ♂ 40 mm., ♀ 45 mm."

The type specimens were received from Dr. Barnes and besides a ♀ co-type there is a series of both sexes in the Barnes collection. The species appears to us to probably represent a local race of *plagiata* Wlk., but for the present we keep it distinct.

HABITAT. Glenwood Springs, Colo., (July 8-30, Aug. 16-23).

O. STYX B. & McD. (Pl. V, Figs. 3, 4).

Olene styx Barnes & McDunnough, Psyche XVIII, 159, Pl. XIII (1911).

"Primaries deep black-brown; all traces of maculation practically lost with the exception of an obscure black basal dash, a large reniform very faintly outlined in whitish and a small white spot above anal angle. Secondaries deep smoky, in the ♀ with outer margin concave below apex, in the male well rounded. Beneath smoky with faint traces of discal spots on both wings. Expanse ♂ 32 mm., ♀ 36 mm.

HABITAT. Duncans, B. C. (Hanham) 1 ♂, 1 ♀. Types, Coll. Barnes."

We are inclined to think this will prove to be a melanic form of *O. plagiata* Wlk. or one of its races. We have an exactly similar ♂ from New Brighton, Pa., (July 2) (Merrick). In the event of our supposition being correct, *styx* would then bear the same relation to *plagiata* that *montana* Beut. does to *manto* Stkr. Breeding however will be necessary to decide the question.

PLATE I.

Fig. 1. *Olene basiflava basiflava* Pack. ♂ ex. 1. Yaphang, L. I. (Coll. Brook. Inst.).

Fig. 2. *Olene basiflava basiflava Pack.* ♀ ex. 1. Yaphang, L. I. (Coll. Doll.).

Fig. 3. *Olene basiflava basiflava* Pack. ♂ New Brighton, Pa. (suffused variety).

Fig. 4. *Olene clintonii* G. & R. ♀ type (Coll. Am. Mus. N. H.).

Fig. 5. *Olene basiflava* Pack. ♂ New Brighton, Pa. (variety) (Coll. Winn).

Fig. 6. *Olene basiflava* Pack. ♀ New Brighton, Pa. (variety).

Fig. 7. *Olene kerrvillei* B. & McD. ♂ Type, Kerrville, Tex.

Fig. 8. *Olene kerrvillei* B. & McD. ♀ Type, Kerrville, Tex.

Fig. 9. *Olene kerrvillei* B. & McD. ♀ Cotype, Kerrville, Tex. (variety).

PLATE I

80

PLATE II.

Fig. 1. *Olene leucophaea* A. & S. ♂ ex. l. Southern Pines, N. C.
Fig. 2. *Olene leucophaea* A. & S. ♀ ex. l. Southern Pines, N. C.
Fig. 3. *Olene basiflava meridionalis* B. & McD. ♂ Type ex. l. Lakeland, Fla.
Fig. 4. *Olene basiflava meridionalis* B. & McD. ♀ Type ex. l. Lakeland, Fla.
Fig. 5. *Olene pini* Dyar ♂ Mass.
Fig. 6. *Olene pini* Dyar ♀ Catskill Mts.
Fig. 7. *Olene plagiata* Wlk. ♂ Hymers, Ont.
Fig. 8. *Olene grisefacta* Dyar ♂ Glenwood Spgs., Colo.
Fig. 9. *Olene grisefacta* Dyar ♀ Glenwood Spgs., Colo. (Agrees with Cotype).

PLATE II

82

PLATE III.

Fig. 1. *Olene vagans vagans* B. & McD. ♂ Type, St. Johns, Que.
Fig. 2. *Olene vagans vagans* B. & McD. ♀ Type, Montreal, Que.
Fig. 3. *Olene vagans grisea* ♂ Cartwright, Man.
Fig. 4. *Olene vagans vagans* ♀ Cotype, Yaphang, L. I. (Coll. Brook. Inst.).
Fig. 5. *Olene vagans grisea* ♂ Type, Eureka, Ut.
Fig. 6. *Olene vagans grisea* ♀ Type, Eureka, Ut.
Fig. 7. *Olene vagans grisea* ♂ Beaver Canon, Ut. (variety) (Coll. Doll).
Fig. 8. *Olene vagans willingi* ♂ Type, Humboldt, Sask.
Fig. 9. *Olene vagans willingi* ♂ Hymers, Ont. (dark variety).

PLATE III

PLATE IV.

Fig. 1. *Olene cinnamomea* G. & R. ♂ Sherborn, Mass.

Fig. 2. *Olene cinnamomea* G. & R. ♀ Biddeford, Me. (suffused variety) (Coll. Winn).

Fig. 3. *Olene cinnamomea* G. & R. ♂ Winchendon, Mass.

Fig. 4. *Olene cinnamomea* G. & R. ♀ Biddeford, Me. (typical) (Coll. Winn.)

Fig. 5. *Olene atomaria* Wlk. ♂ (Coll. Doll).

Fig. 6. *Olene atomaria* Wlk. ♀ Lititz, Pa. (Coll. Doll).

Fig. 7. *Olene atomaria* form. *parallela* G. & R. ♂ New Brighton, Pa.

Fig. 8. *Olene atomaria* form. *parallela* G. & R. ♀ Catskill Mts. (Coll. Field Mus.).

Fig. 9. *Olene atomaria* form. *parallela* G. & R. ♂ New Brighton, Pa. (variety).

PLATE IV

PLATE V.

Fig. 1. *Olene manto manto* Stkr. ♂ ex. l. Southern Pines, N. C.
Fig. 2. *Olene manto manto* Stkr. ♀ ex. l. Southern Pines, N. C.
Fig. 3. *Olene styx* B. & McD. ♂ Duncans, Vanc. Is.
Fig. 4. *Olene styx* B. & McD. ♀. Duncans, Vanc. Is.
Fig. 5. *Olene vagans willingi* B. & McD. Larva.
Fig. 6. *Olene vagans vagans* B. & McD. Larva?
Fig. 7. *Olene basiflava meridionalis* B. & McD. Larva.

PLATE V

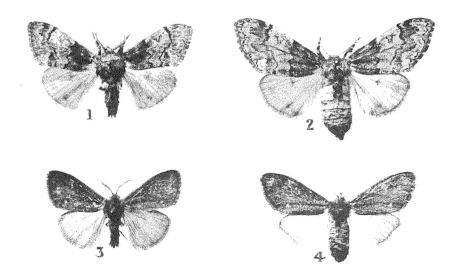

PLATE VI.

Fig. 1. *Parorgyia obliquata* G. & R.   Type ♀ "Grote & Robinson Coll."
Fig. 2. *Parorgyia obliquata* G. & R.   Type ♀ "Grote & Robinson Coll. E. S."
(Eastern States).
Fig. 3. *Parorgyia parallela* G. & R.   Type ♀ "Grote & Robinson Coll."
Fig. 4. *Parorgyia parallela* G. & R.   Type ♂ "Grote & Robinson Coll."
So called ♂ type but spurious as *parallela* was described from a single ♀.
Fig. 5. *Parorgyia cinnamomea* G. & R.   Type ♂ "Grote & Robinson Coll. Br."
Fig. 6. *Parorgyia cinnamomea* G. & R.   Type ♂ "Grote & Robinson Coll. E. S."
Fig. 7. *Parorgyia clintonii* G. & R.   Type ♂ "Grote & Robinson Coll. E. S."
Fig. 8. *Olene montana* Beut.   Type ♀ "Summit Black Mts., N. C., 6500 ft."
Fig. 9. *Olene montana* Beut.   Type ♂ "Summit Black Mts., N. C., 6500 ft."
The above types are all in the Coll. of the American Museum of Nat.
History, New York.

PLATE VI

# PLATE VI

PLATE VII

Fig. 1. *Acyphas plagiata* Wlk. Type ♂. Brit. Mus.
Fig. 2. *Dasychira atomaria* Wlk. Type ♀. Brit. Mus.
Fig. 3. *Edema plagiata* Wlk. Type ♀ "Orillia, Ont." Brit. Mus.
Fig. 4. *Olene manto* Stkr. Type ♂ "Stewart Co., Ga." Field Mus.
Fig. 5. *Olene atrivenosa* Palm. Type ♂. Coll. Palm.
Fig. 6. *Olene atrivenosa* Palm. Type ♀. Coll. Palm.

INDEX

achatina A. & S.........47, 48, 49, 53

atomaria Wlk........47, 48, 49, 52, 67

atrivenosa Palm.........47, 48, 49, 65

basiflava Pack.....47, 48, 49, 52, 54, 55

cinnamomea G. & R.....47, 48, 49, 70

clintonii G. & R.....47, 48, 49, 54, 55

v. grisea B. & McD.......... 63

grisefacta Dyar............. 47, 75

v. interposita Dyar........... 47, 72

kerrvillei B. & McD........ 59

leucophaea A. & S. ...47, 48, 49, 52, 65

leucostigma A. & S.......... 50

manto Stkr.............47, 49, 71, 72

v. meridionalis B. & McD.... 58

mendosa Hbn.............. 52

montana Beut.............. 47, 72

obliquata G. & R......47, 48, 49, 52, 67

parallela G. & R.........47, 48, 49, 69

pini Dyar.................. 47, 74

v. *pinicola* Dyar 47, 74

plagiata Wlk. (Acyphas)...47, 48, 49

plagiata Wlk................ 50, 74

Styx B. & McD............. 47, 76

tephra Hbn..............47, 48, 49, 60

vagans B. & McD............ 60, 61

v. willingi B. & McD......... 64

TO THE

NATURAL HISTORY

OF THE

LEPIDOPTERA

OF

NORTH AMERICA

VOL. II
No. 3

NEW N. AM. LEPIDOPTERA
with
NOTES ON DESCRIBED SPECIES

BY

WILLIAM BARNES, S. B., M. D.

AND

J. H. McDUNNOUGH, Ph. D.

DECATUR, ILL.
THE REVIEW PRESS
APRIL 15, 1913

Published
Under the Patronage
of
MISS JESSIE D. GILLETT
Elkhart, Ill.

DIURNALS

Argynnis eurynome Edw.

We have recently had the opportunity of examining the specimens of this species and its numerous so-called races or varieties contained in the W. H. Edwards Collection in Pittsburg and have made the following notes thereon:

eurynome Edw. (Tr. Am. Ent. Soc. IV, 66, 1872).

Described from specimens taken in Colo. by Mr. Mead; in the Edwards Coll. are ♂ and ♀ types labelled "Colo." The name will apply to the *silvered* form with pale yellow ground color on secondaries beneath, with more or less distinct greenish shading at base and slight cinnamon brown suffusion over the disk; the ♀'s are paler, with contrasting yellow subterminal spots and blackish terminal border above and with stronger suffusion of green below. This form is typical in the higher mountain regions of Colorado (Hall Valley, Silverton) and we presume it extends into the neighboring states. Edwards' figures (Butt. II, Pl. 23) show rather more green on underside than is usual. Occasional unsilvered specimens occur in Colo. but these represent mere aberrations and should not be confused with races where the lack of silver on the spots is characteristic. *Macaria* Edw. listed by Dyar as a synonym and described from Havilah, Calif., does not even belong in this group.

erinna Edw. (Can. Ent. XV, 33, 1883).

Described from Spokane Falls, Wash. Specimens so labelled are marked "type" in the Edwards Coll. The race is characterized by reduction of black on upper side and rather brighter coloration, especially in ♀ ; on underside of secondaries the spots are silvered, all trace of green is wanting, and the disk is suffused to a considerable extent with cinnamon-brown. We have specimens from Ft. Klamath, Oregon, which agree exactly with the types; it is evidently, to judge from the localities, a form of the lower mountain regions of the Pacific Coast States; *arge* Strecker is apparently closely related and may prove identical.

CLIO Edw. (Tr. Am. Ent. Soc. V, 106, 1874).

Described from ♂'s taken in the Teton Mts., Montana, by the Hayden expedition and a ♀ taken by Dr. Coues of the North Boundary Line Expedition. The types in the Edwards Coll. are labelled respectively "B. Am. (Geddes)" and "Head of Sask. Brit. Am. (Bruner)" and are thus *undoubtedly spurious;* we did not see the true types. The specimens so labelled in the Collection are so close to what Edwards later described as *artonis* that we were unable to find any satisfactory point of distinction; both are characterized by *unsilvered* spots on underside with a great tendency towards obsolescence of the maculation; the ground color of secondaries is pale yellowish with a slight tinge of green towards base and a faint washing of cinnamon-brown on the disk. Before definitely placing *artonis* as a synonym it will be necessary, if possible, to trace the authentic types.

ARTONIS Edw. (Tr. Am. Ent. Soc. IX, 1, 1881).

Described from 1 ♂ from Big Horn, Montana, and 2 ♀'s from Wells, Elko Co., Nevada. The types (♂ and ♀) in the Edwards Coll. are marked Elko, Nevada; the ♂ type from Montana we did not find. The type should be restricted to this Montana ♂ in case the two forms prove distinct; in this case there is every probability of *artonis* proving a synonym of *clio,* both types coming from the same general locality. We have specimens from Park City, Utah, which agree exactly with the Nevada types in lack of silver spots and general obsolescence of markings on underside of secondaries.

OPIS Edw. (Tr. Am. Ent. Soc. V, 105, 1874).

Described from specimens taken at Bald Mt. Cariboo, B. C., by Mr. Crotch. 2 ♂'s and 1 ♀ from this locality marked 'type' are present in the Edwards Collection; the species is also figured in Butt. II, Pl. 25. It represents a small *unsilvered* race with rather heavy black markings on upper side and considerable green scaling at base and over disk of underside of secondaries, the spots are distinct. It is not synonymous with *clio,* as listed by Dyar, but shows much greater affinity to the following subspecies *bischoffi* Edw.

BISCHOFFI Edw. (Tr. Am. Ent. Soc. III, 189, (1870).

Described from 1 ♂ 1 ♀ taken in Alaska opposite Kodiak by Mr. Bischoff and received from Mr. Behrens. In the Edwards Coll. four specimens are labelled 'type', one pair from Kodiak and a second pair from Sitka; these latter must be considered as spurious. The original description calls for an *unsilvered* form, a feature borne out by the Kodiak types, but Edwards in his Butt. II, Pl. 25 figures a silver marked species, evidently one of the Sitka specimens, and remarks that the species occurs around Sitka in both silvered and unsilvered forms. The name, *in sens. strict.* can only apply to the unsilvered form; as stated above the race is very close to *opis,* but has heavier and more extended black shading at base of wings on upperside and the green of underside is more bronze-colored and diffuse. We do not know why Dyar has listed this as a variety of *eurynome;* according to rules of nomenclature *eurynome* might become a variety of *bischoffi,* but never *bischoffi* of *eurynome,* as the name has two years priority.

In conclusion we venture to describe two new races of which we have long series of both sexes and which are sufficiently distinct from described forms to warrant a name in our estimation.

A. BISCHOFFI WASHINGTONIA subsp. nov. (Pl. I, Figs. 5-8).

Upper side of both sexes very similar to Colorado *eurynome,* with rather more black dusting at base of wings, but less so than in typical *bischoffi;* size considerably smaller than the average *eurynome.* Underside with the spots *well silvered;* ground color of secondaries a clear pale yellow with basal area of wings to post-median row of spots heavily and evenly scaled with green, leaving only traces of ground color in the cell. Expanse ♂ 43 mm., ♀ 47 mm.

HABITAT Mt. Ranier, Wash. (7000 ft.) (July 24-30) (McDunnough). 7♂, 7♀. Types, Coll. Barnes.

We consider this to be the southern representative of the Alaska form; it is probably found on all the higher peaks of the Coast Range through Washington and Oregon; Edwards in his description of *erinna* mentions a similar ♀ from Mt. Hood. In a series of over 50 specimens captured only a single unsilvered aberration was found, so that the silvered spots may be considered an attribute of the race. *Opis* probably represents a British Columbia form of the same species in which apparently the silvering has been lost; more material from this region and from Alaska is necessary however to definitely define the races.

A. EURYNOME LUSKI subsp. nov. (Pl. I, Figs. 1-4).

♂ Upper side as in typical *eurynome* from Colorado; beneath, primaries rosy, pale ochreous along costa and at apex with the veins bordered terminally with orange-brown; subterminal spots pale, unsilvered; secondaries with the spots entirely unsilvered, pale ochreous, washed with green along costa, at base of wing, and slightly over disk; the veins, the space between the postmedian series of spots, the basal interspaces and the terminal area suffused with orange-brown.

♀ Paler than ♂, much as *eurynome* ♀, with pale subterminal spots and rather heavy black terminal border on upper side; beneath primaries deeper rosy, subterminal spots shaded with green towards apex; secondaries much more suffused with green generally, subterminal row of spots rarely with traces of silver scaling, other spots unsilvered; orange-brown shading often prominent beyond the postmedian spots, encroaching on the pale band. Expanse ♂ 47 mm. ♀ 51 mm.

HABITAT. White Mts., Arizona (Lusk). 7 ♂, 7 ♀. Types, Coll. Barnes.

Evidently a very distinct race, characterized by the unsilvered underside of secondaries and the peculiar contrasted appearance, due to the mixture of green and orange-brown scaling. We take pleasure in naming it after the collector, Mr. R. D. Lusk, who has brought many interesting forms to light as a result of two seasons' collecting in the White Mts., Arizona. We prefer to consider this a form of *eurynome* rather than *bischoffi* until it has been definitely proved that the two names represent mere racial forms.

BRENTHIS CHARICLEA V. BOISDUVALI Dup. (Pl. II, Figs. 5, 6).

In Grönlands Insekt-fauna (1890) Aurivillius figures typical *chariclea* Schneid. *arctica* Zett. and var. *boisduvali* Dup. restricting the latter name to the Labrador race; this race differs from the type form from Lappland and Finnmark in the heavier purple suffusion on underside of secondaries and in the fact that the central white median band of spots is largely suffused with reddish-brown, leaving only portions of the white color visible. This race extends across the northern portion of the continent without any marked degree of variation; we have specimens from Labrador; Hymers, Ont.; Calgary, Alberta; Saskatchewan, before us which cannot be differentiated into well-defined races. Two years ago we captured a long series of specimens on Mt. Ranier, Washington which shows sufficient constant points of distinction from our other specimens to warrant a name; we characterize the race as follows:

B. CHARICLEA RAINIERI subsp. nov. (Pl. II, Figs. 1-4).

♂ Upper side as in typical *boisduvali* with a slight reduction of the black markings as compared with Labrador specimens. Beneath, secondaries much more contrasting in coloration, median band with less red-brown suffusion, often almost entirely pale yellow, the black border to this band not so heavy and the spot in the 4th interspace from costa more drawn out, being intermediate in this respect between *boisduvali* and typical *arctica* Zett. from Greenland; between veins 2-5 terminally a large patch of pale yellow extending inwardly beyond the subterminal row of round spots, which are scarcely as prominent as in *boisduvali* and are surrounded by more or less pale yellow suffusion.

♀ Very characteristic on upper side, paler than the ♂, the wing suffused apically and along costa with greenish yellow; terminal row of spots pale yellow, not concolorous as in typical *boisduvali* ♀'s, which also lack the pale yellow suffusion, presenting no contrasts; underside of secondaries much as in ♂ but the yellow suffusion is usually greater and has a distinct greenish tinge; the costal spot of the median band is usually white as well as the rather large terminal lunules.

HABITAT. Mt. Ranier, Wash. (July 24-31) (6-7000 ft.) (McDunnough), 7♂, 7♀. Types, Coll. Barnes.

The species was the commonest butterfly collected, being found all over the grassy slopes around the so-called Paradise Valley. It reminds one in the greenish yellow suffusion of the ♀ of the var. *isis* of the European *pales*. Dr. Dyar evidently refers to this species in Proc. Wash. Ent. Soc. V, 129 under the name var. *arctica* Zett.

BRENTHIS TRICLARIS Hbn.

This species as figured by Hübner (Samml. Ex. Schmett. II, Pl. 232) shows very distinct silver markings on the underside; Moeschler (Wien. Ent. Monatsch, IV, 337) applied the name to Labrador specimens, considering them a racial form of the European *aphirape* Ochs. and making *ossianus* Bdv. a synonym. He has been following in this by the majority of later authorities. The Labrador race, to judge by a small series before us, shows considerable tendency to lose the silver on the underside, especially in the ♂ sex; five specimens from Hymers, Ont., and one from Saskatchewan have however very prominent silver markings and in this respect approach Hübner's figure more closely than does the Labrador form; our series are however not nearly lengthy enough to decide how great the individual variation from each locality may be.

On the strength of Mead's report (Rep. Wheel. Ex. 5, 756) the species has been listed from Colorado; this Colorado form, which is quite constant, as far as can be judged from a good series of both sexes before us, is noticeably different from our Labrador and Canadian specimens of *triclaris;* The upper side is paler with the black basal shading reduced and the black markings narrow and cleanly cut; the underside of secondaries resembles that of the type form *aphirape* greatly, in that the rows and bands of spots are all pale yellow without any silver except a faint trace at times on the marginal lunules; the ground color is however deep cinnamon-brown and not yellowish as in *aphirape*. We would propose the name APHIRAPE ALTICOLA subsp. nov. for this Colorado form and have made types and cotypes of 5 ♂'s and 2 ♀'s from Hall Valley, Colo., (June 21-30) (Barnes) contained in Coll. Barnes.

CHLORIPPE MONTIS Edw.

Dr. Skinner in his paper on the Boreal Am. Species of the genus *Chlorippe* says of this form "*Montis* is only slightly different from the Texan *antonia* and is found in Colorado. It is lighter in color in some specimens and is only a mountain mutation." If Dr. Skinner had consulted the original description (Pap. III, 7) he would have discovered that the race was described in a paper entitled "Notes on the collection of butterflies made by Mr. H. K. Morrison in Arizona, 1882" and that, although a single ♂ taken by Mr. Dodge at Boulder, Colo., is mentioned as belonging to this variety, the description of the species was drawn up from 3 ♂, 3 ♀ from Mt. Graham, Arizona, which must be regarded as the type locality.

On a recent visit to the Edwards Collection we were unable to find any of these types; a series labelled v. *montis* by Edwards (apparently erroneously) consisted of Colorado specimens which scarcely differed from *antonia,* and certainly not sufficiently to warrant a name; one specimen was found labelled "*Antonia* ♂, Dodge, type of *montana*" which is evidently the specimen referred to by Edwards in the course of his description.

The original description calls for an entirely different insect to the Colorado form; Edwards states "The Arizona examples are *fulvous* above, *bluish gray* beneath but in other points they agree with *antonia*". Later in the description he calls attention to the lack of white on the underside of secondaries.

We have a long series of both sexes from Arizona which agree exactly with this description and the racial name appears to us very well grounded. The form is at once separated from *antonia* by the bright fulvous appearance of the upper side; in coloration it approaches very closely *leilia* but is separated from this form by the fact that there are *two* spots in the cell of primaries towards base of wing superimposed but not coalescing to form a bar as in *leilia;* the white bar in the cell is also lacking. Hollands figure (Pl. XXIII, Fig. 11) referred to by Skinner under *leilia* represents this species. Regarding *A. cocles* Lint. which Dr. Skinner regards as the ♀ of *antonia* we would call attention to the fact that the original description states distinctly "discoidal cell *double-barred,* the bars black-bordered and ochraceous within; *A. celtis, A. alicia* and *A. antonia* have a single bar outwardly and towards the base *two separate spots*—more contiguous in the latter species." We have not seen the type, but judging by the above remarks should incline to place *cocles* as a synonym of *leilia* following Aaron and Coolidge; we have a ♀ from the neighborhood of Brownsville, Texas, under this name which only differs from Arizona specimens in the slightly duller color.

We have recently received from Arizona a form of *clyton* allied to *texana* Skin. but sufficiently distinct to warrant a name; it bears the same relation to *clyton* that *montis* does to *celtis*.

C. CLYTON SUBPALLIDA subsp. nov. (Pl. II, Figs. 7-9).

Upper side much as in *texana* Skin., bright fulvous deepening into purple-brown towards apex of primaries; spots on primaries as in type form, pale creamy, those of the subterminal row rather reduced and the one in the interspace between veins 2 and 3 showing traces at times of a fulvous ring; secondaries without black shading, fulvous with the usual subterminal row of black ocelli, the brown rings merging more or less into the ground color of the wing. Beneath with the secondaries heavily washed with ochreous gray, leaving the maculation very indistinct and not contrasting; nearly all trace of central white shading wanting; primaries brighter ochreous at base, shading into purplish-gray outwardly with fairly well defined maculation. Expanse ♂ 47 mm., ♀ 67 mm.

HABITAT. Babaquivera Mts., Pima Co., Arizona (July-Aug.) 2 ♂'s, 4 ♀'s. Types, Coll. Barnes.

The chief point of distinction from *texana* Skin. (Pl. II, Fig. 10) is the suffused underside of secondaries and indistinct maculation.

HESPERIDAE

Copaeodes rayata sp. nov. (Pl. III, Figs. 1, 2).

♂ Antennae orange, paler beneath, shaded with black above; palpi, head and abdomen above orange, beneath whitish; primaries bright yellow-orange, costa slightly black apically; sex-mark a long fine black streak just below and parallel to median vein; a faint terminal dark line; secondaries orange with black scaling along costa and at base of wing; fringes of both wings concolorous, paler outwardly, smoky at apex of primaries. Beneath slightly paler orange than above, primaries shaded at base with black which juts out as a streak on median vein; at apex with slight whitish shading; secondaries with a prominent pale whitish ray extending from base of wing through cell to outer margin, interspaces of other veins slightly rayed with white.

♀ Primaries with median vein broadly black to end of cell, the color extending finely along veins 2, 3, and 4; anal vein outlined in black; other veins at times also black; secondaries black at costa and base, this color tending to extend along the veins, especially along anal vein; a black terminal line to both wings; fringes more or less suffused with smoky, especially apical half of primaries. Expanse 18-20 mm.

Habitat. San Benito, Texas (July 16-23) (Dorner) 7 ♂, 3 ♀. Types, Coll. Barnes.

In the W. H. Edwards Coll. 3 ♀'s from Texas of this species are labelled *"arene* Edw." in Edwards own hand. These cannot however be considered types as *arene* was described originally (Tr. Am. Ent. Soc. III, 214) from a specimen from Ariz. which is not in the Edwards Coll. It is probably due to this error of identification that Edwards later redescribed the true *arene* under the name of *myrtis*, as pointed out by Dyar (Jour. N. Y. Ent. Soc. XIII, 126). The white dash on under side of secondaries renders our new species very easily recognizable.

ARCTIADAE

CRAMBIDIA PURA sp. nov. (Pl. IV, Figs. 5, 6).

Head, thorax, abdomen, and primaries pure white, immaculate. Beneath, primaries and costal half of secondaries pale smoky, remainder of secondaries white. Expanse ♂ 20 mm., ♀ 22 mm.

HABITAT. Southern Pines, N. C. (Sept. 24-30) (A. Manee). 1 ♂, 1 ♀. Types, Coll. Barnes.

The species resembles *casta* Sanb. in coloration, is however scarcely half the size; *casta* is a northern species ranging from the N. E. States across the continent into British Columbia; our species is evidently southern in its range.

CRAMBIDIA IMPURA sp. nov. (Pl. IV, Fig. 4).

Head and thorax whitish suffused with gray, abdomen clothed with whitish hairs; primaries white, more or less tinged with smoky; secondaries smoky-brown with paler fringes. Beneath, uniform smoky with apical half of costal margin and fringes pale whitish. Expanse 32 mm.

HABITAT. Palmerlee, Arizona. (Sept. 8-15) (Biederman); White Mts., Arizona. (Lusk). 8 ♂'s. Type, Coll. Barnes.

The species closely resembles *suffusa* B. & McD. from S. Calif., but lacks all traces of yellow on front characteristic of both *suffusa* and *cephalica* G. & R. As we have seen no Arizona specimens with yellow head we consider that we are dealing with at least a well marked race. Our White Mts. specimens are slightly more ochreous in tinge than those from Palmerlee.

CRAMBIDIA DUSCA sp. nov. (Pl. IV, Fig. 7).

Head and front deep gray, thorax paler; primaries gray with an ochreous tinge, dusted in the interspaces with smoky scales leaving the veins very slightly paler; secondaries deeper smoky brown. Beneath uniform smoky brown, slightly paler along costa of primaries. Expanse 26 mm.

HABITAT. San Diego, Calif. 23 May, 25 Oct. (Wright) 27 May (Field). 3 ♂. Type, Coll. Barnes. Cotypes with Messrs. Field and Wright.

Closely resembles *pallida* Pack. in coloration, but the areole is present on primaries and the veins are not so distinctly outlined. The lack of yellow front separates it from *suffusa* B. & McD. than which it is smaller and darker. We have two apparently similar specimens from Brownsville, Texas.

OZODANIA SUBRUFA sp. nov. (Pl. III, Figs. 4, 5).

Front brown, palpi, head, and thorax pale ochreous, abdomen pink; primaries gray-brown more or less suffused with ochreous, giving a very characteristic mottled appearance; a pale ochreous basal patch on inner margin not connected with a triangular patch of similar color just before anal angle; this latter patch may or may not be joined by a perpendicular narrow ochreous line to a smaller costal patch of same color; the brown ground color of wing is usually deepest on the margins of these light patches. Secondaries deep rose somewhat paler along costa with a small smoky patch at apex of wing only. Beneath rose-color, costa of primaries very narrowly blackish to transverse ochreous band which is faintly visible; beyond this band the apex of wing is broadly smoky, this color extending narrowly along outer margin to anal angle; secondaries rose with the faintest trace of smoky at apex. Legs ochreous, banded with black. Expanse ♂ 13 mm.; ♀ 16 mm.

HABITAT. San Benito, Texas, (July 8-31, Aug. 1-7) (G. Dorner) 2 ♂'s, 5 ♀'s. Types, Coll. Barnes.

We retain the generic term *Ozodania* Dyar, which has been sunk as a subgenus of *Illice* by Hampson, as our specimens show in the male sex the characteristic tuft of hairs on underside of tornus on primaries and a strong anal lobe on secondaries with tufts of hair. The species is characterized by the mottled appearance of primaries and the rose-colored underside; it is possible that the Mexican specimen mentioned by Dyar in his description of *schwarziorum* (Psyche VIII, 359) may belong to this species, in which case the name *schwarziorum* should be restricted to the Arizona form; this has the underside, with the exception of the median band, mouse gray.

Our new species was already figured in Vol. II, No. 1 of our Contributions.

ILLICE CONJUNCTA sp. nov. (Pl. IX, Fig. 14).

Palpi smoky, tipped with pale ochreous, front smoky; head and thorax whitish, base of collar slightly pink; abdomen pink; primaries deep smoky with large semiquadrate white spot on costa beyond middle and a broad white band along inner margin enlarging before anal angle into a white patch which is slightly defined towards base of wing by some fuscous scales on inner margin; secondaries pink with a large marginal smoky patch occupying nearly half the wing; the inner margin of this patch extends obliquely inwards from apex of wing across cell to near origin of vein 2, then curves rapidly outwards, attaining outer margin considerably before anal angle; fringes, except at anal angle, dusky. Beneath as above, but spots and band of primaries slightly suffused with pink. Expanse 15 mm.

HABITAT. San Benito, Texas (G. Dorner) 1 ♂. Type, Coll. Barnes.

Allied to *subjecta* Wlk. but the maculation is paler and the whole area between subcostal fold and inner margin is filled with the white band.

PYGARCTIA FLAVIDORSALIS sp. nov. (Pl. IV, Fig. 3).

Antennae black; palpi gray-brown, yellow basally; legs gray, coxae of fore legs yellow, hind femora and tibia paling into whitish; front and head and abdomen dorsally yellow, latter with dorsal row of black spots; thorax, patagia and wings white, immaculate. Beneath, abdomen and wings white. Expanse ♂ 30 mm., ♀ 32 mm.

HABITAT. San Benito, Texas (Mch. 16-23) (July 16-23) (Aug. 1-7) (G. Dorner) 4♂, 3♀. Types, Coll. Barnes.

The species is close to *elegans* Stretch., differs from it in the abdomen and front being distinctly yellow and not rosy. As this feature is constant in all the specimens before us we conclude that if not a good species it is at least a distinct race and worthy of a name.

PYGARCTIA MURINA ALBISTRIGATA subsp. nov. (Pl. III, Figs. 10, 12).

Head, thorax and primaries mouse-gray, the latter with a distinct whitish spot or mark on the discocellulars and usually with the veins beyond the cell streaked with the same color. Secondaries in the ♂ whitish with broad smoky border, in ♀ entirely smoky. Abdomen, collar, and edges of patagia scarlet. Expanse ♂ 25 mm., ♀ 28 mm.

HABITAT. San Benito, Texas (May-June) (G. Dorner) 6♂'s, 3♀'s. Types, Coll. Barnes.

Differs from the type form from Arizona in the pale striations of primaries. Hampson places *murina* in the genus *Euchaetes* but a long series of specimens in Coll. Barnes all show the claw on front tibia characteristic of *Pygarctia*. Our Texan race also possesses the claw.

NOCTUIDAE

Schinia brunnea sp. nov. (Pl. III, Figs. 7-9).

Palpi, head, and thorax brown; primaries brown, shaded with pale ochreous and slightly tinged with olivaceous; basal space to t. a. line rather even brown with a very slight admixture of pale scaling; t. a. line formed by the outer margin of this basal space, but slightly darker than the space itself, bent outward from costa to below cell then almost perpendicular to inner margin; inner half of median space next t. a. line pale ochreous, the palest portion of the wing, outer half shaded heavily with brown with traces of a diffuse broad median shade, most noticeable as a dark patch on costa above reniform; orbicular oval, olivaceous, with slight traces of black marginal line; claviform similar to orbicular, resting on t. a. line; reniform semiquadrate, outlined laterally with black; t. p. line evenly sinuate, indistinct at costa, outlined in deep brown above inner margin; s. t. line pale, irregular, angled below costa, with strong inward angle on vein 5 almost reaching t. p. line, excurved opposite cell and again more strongly between veins 2 and 5, narrowing at inner margin; subterminal area brown with slight bluish scaling in the bulge between veins 2 and 5; terminal space paler; a broken black terminal line; fringes brown, cut with blackish outwardly. Secondaries pale ochreous with a broad black border, suffused outwardly with brownish; prominent black discal spot; fringes pale; slight black scaling at base and along inner margin. Beneath, primaries ochreous shaded with brownish along costa and outwardly; orbicular, reniform and claviform prominently black, a diffuse black shade above the tornus, traces of t. p. and s. t. lines at costa; terminal line and fringes as above; secondaries ochreous shaded with pinkish brown along costa and outer margin, with prominent black discal spot and dark shade patch on outer margin near tornus; one or two black streaks on inner margin near base of wing. Expanse 23 mm.

Habitat. Loma Linda, S. Bern. Co., Calif. (Oct. 8-15) (Pilate); San Diego, Calif. (Ricksecker) 16 ♂'s, 8 ♀'s. Types, Coll. Barnes.

The species approaches *tertia* Grt. in general type of maculation. It is extremely variable; some specimens before us are so suffused with brown as to render all the maculation very indistinct, others, notably ♀'s, are more contrasting in coloration than our types, the brown being deeper, especially in subterminal area where the bluish scaling at times is quite prominent; the brown color may vary from pale ochreous to deep olivaceous and the maculation of the under side may be very distinct or almost wanting except the black discal spots. Our types represent specimens midway between the two extremes.

INCITA AURANTIACA Edw. (Pl. III, Fig. 6).

Typical specimens from California (we have a ♂ before us that has been compared with type) have dark primaries, and secondaries with a moderate black border and prominent discal spot. We agree with Smith (Jour. N. Y. Ent. Soc. XV, 141) that *Pyrocleptria californica* Hamp. is a synonym. We have specimens from Arizona which evidently form a distinct race of this species which we characterize as follows:

I. AURANTIACA TENUIMARGO subsp. nov. (Pl. III, Fig. 3).

Primaries paler than in type form, median and terminal areas largely pale ochreous; secondaries with the outer border reduced to a mere line between vein 2 and anal angle, broadening slightly towards costa, discal spot smaller and rather indistinct.

HABITAT. Redington, Ariz.; S. Ariz. 1♂, 2♀'s. Types, Coll. Barnes.

We have several more specimens in Coll. Barnes which are too poor to make types, but which agree in markings with the above.

GROTELLA SPALDINGI sp. nov. (Pl. IV, Figs. 1, 2).

Head, thorax, and primaries pale creamy to ochreous; maculation very variable, in well marked specimens the t. a. line is represented by a dark spot in the cell, another in submedian fold and some scaling on inner margin just before middle of same, these may be more or less connected by a line of similar color proceeding slightly outwardly oblique from costa to inner margin, or the maculation may be reduced to the spot in submedian fold; t. p. line, when present, slender, brown, outcurved from costa to beyond cell, then inwardly oblique to a point on costa close to t. a. line; this line is rarely complete, and in many cases is reduced to a spot beyond the cell, one in the submedian fold and slight scaling on inner margin. Secondaries smoky with paler fringes. Beneath, primaries smoky, with ochreous costa and outer margin, secondaries pale ochreous. Expanse 20 mm.

. HABITAT. Vineyard, Utah (June-July) (T. Spalding). 10♂'s, 2♀'s. Types, Coll. Barnes.

This species differs from typical *Grotella* species in the shape of the central process to frontal prominence; in *Grotella* this is truncate, in our species it forms a very prominent knife-like vertical ridge; the mid-tibiae are spined and the anterior tibiae armed with two slender claws at extremity. As the general appearance is distinctly like *Grotella* we prefer to keep the species for the present in this genus rather than create a new one based on slight differences of the frontal prominences. We take pleasure in naming the species after Mr. Spalding, the well-known collector.

POLIA (MAMESTRA) APURPURA sp. nov. (Pl. VII, Fig. 2).

Palpi and front very roughly haired, dark gray-brown shaded with black; collar dark gray crossed by a blackish median line and tipped with whitish; thorax and patagia dark gray-brown, grizzled; abdomen smoky-ochreous; antennae in ♂ strongly serrate and fasciculate; primaries very deep even gray-brown with maculation clear-cut and contrasting; a black dentate basal ½ line followed by a pale gray shade-line; t. a. line geminate, inner line indistinct, outer line blackish, filled with light gray, very strongly dentate in submedian fold and above inner margin, in general course perpendicular to inner margin; orbicular large, oval, touching t. a. line, filled with pale gray and slightly outlined with blackish; claviform small, outlined in black, resting on submedian tooth of t. a. line; reniform large, lunate, annulate with pale gray and black and filled with a gray shade very little lighter than ground color of wing, followed to t. p. line by slight ochreous shading; median shade very faint, slightly visible in costal portion of wing; t. p. line single, blackish, starting from middle of costa and strongly bent out around reniform, then parallel to outer margin and prominently dentate on the veins, followed by a faint pale gray shade line, representing the filling of a geminate line in which the outer line is obsolete; s. t. line pale, very irregular, preceded by black-brown velvety shading, the most prominent portion of the maculation, oblique from costa to vein 7, with slight tooth on vein 8, then bent out strongly towards margin and again oblique and slightly irregular to below vein 5, forming on veins 3 and 4 a prominent W mark which almost touches outer margin, oblique again to submedian fold and then bent out to anal angle, the black shading most prominent below the W mark; terminal lunulate dark line and pale line at base of fringes; fringes dark, slightly cut by pale ochreous opposite the veins. Secondaries creamy, shaded with smoky, especially strongly in subterminal area; traces of a diffuse postmedian line; dark terminal line; fringes smoky, paler towards anal angle. Beneath smoky, paler towards base, both wings crossed by a broad smoky postmedian line, marked more strongly on the veins. Expanse 46 mm.

HABITAT. White Mts., Ariz. (Lusk) 5 ♂. Type, Coll. Barnes.

This species belongs in the *purpurissata* group, but may be distinguished by its much deeper coloration and the much more prominent W mark of subterminal line.

POLIA (MAMESTRA) LUSKI sp. nov. (Pl. VII, Fig. 8).

Collar gray tipped with brown and crossed by a dark median line; thorax gray-brown, abdomen ochreous; primaries brown, suffused with light purple-gray, costa light ochreous at base, a slight black basal dash below median vein and a black angular mark above inner margin near base preceded by light ochreous scaling; t. a. line black, outwardly inclined, geminate, inner line faint, filled with pale purple-gray shade, angled three times, in cell, in submedian fold and below vein 1; orbicular oval, oblique, outlined with black, open towards costa, filled with pale gray; claviform large, black, extending ½ across median space, reniform upright, large, hollowed out centrally on outer side, annulate with pale ochreous and black, more or less open towards costa, filled with light brown which shades into smoky towards base; slight outwardly bent median shade, most distinct below reniform; veins 1 and 2 in median space sprinkled with whitish; t. p. line bent out below costa and slightly incurved in submedian fold, finely scalloped, followed by a pale shade line; subterminal space shaded with purple-gray; s. t. line prominent, yellow, indistinct at costa where it is defined by the contrast between dark subterminal costal patch and pale gray apical patch, bent sharply out at vein 7, then very even and parallel to outer margin with slight inward angle below vein 2; preceded by brown shading; terminal dark brown lunules and dusky fringes. Secondaries smoky, with darker terminal area and faint yellow terminal line, broadening below vein 2. Beneath smoky, apex of primaries and terminal border of secondaries somewhat darker; indistinct discal dots. Expanse 34 mm.

HABITAT. White Mts., Ariz. 2♂. Type, Coll. Barnes.

Belongs close to *dodi* Sm. but differs in the browner coloration and the entire lack of W mark to subterminal line. We take pleasure in naming the species after the collector, Mr. E. Lusk.

POLIA (MAMESTRA) BICOLOR sp. nov.

Head and thorax whitish gray, patagia heavily tipped and sprinkled with black; primaries largely blackish, suffused in places with white, maculation rather diffuse and indistinct; the ordinary lines are geminate, filled with white, which on account of the dark ground color alone shows clearly in most cases; subbasal line distinct, angulate; t. a. line outwardly inclined, dentate; orbicular outlined in white on dark ground, round; reniform large, more or less filled with white, with a darker lunate mark centrally; both spots preceded by white shading on costa; t. p. line dentate, rather evenly sinuate, followed above inner margin by a large patch of whitish, extending to s. t. line, semicircular in form; s. t. line indistinct, white, irregular, approaching outer margin on veins 3 and 4 and slightly dentate, preceded by faint white scaling; fringes pale, checkered with smoky, more prominently in basal half. Secondaries in ♂ white, with apical smoky shading and dark terminal line; in ♀ deep smoky, with pale fringes cut by a dark subbasal line not attaining anal angle. Beneath, in ♂ white, primaries heavily smoky in costal half of wing except costa itself which is narrowly white, powdered with smoky; secondaries slightly dusted with smoky along costa with small discal spot; terminal dark line to both wings;. in ♀ whole of primaries smoky with slightly paler costa and traces of subterminal line; secondaries white, heavily sprinkled with smoky, especially apically, with discal dot and subterminal line. Expanse 29 mm.

HABITAT. Kerrville, Tex. (Oct.) (Lacey) 2♂, 3♀. Types. Coll. Barnes.

The species is allied to *palilis* Harv. but does not show the sexual dimorphism found in this species and is generally a much darker insect, it was figured in Vo\'l. II, No. 1 of our Contributions.

ERIOPYGA EUXOIFORMIS sp. nov. (Pl. V, Fig. 3).

♂. Antennae ciliate; head, thorax and primaries purple-brown; the lines are geminate, but only one shows distinctly and the pale filling thus presents the appearance of the line itself; basal line upright, extending not quite across wing, represented by a pale ochreous shade and a dark outer line; t. a. line rather indistinct, outwardly oblique, crenulate, composed of ochreous shade and outer dark line; on this rests a small loop-like claviform filled with ochreous; orbicular oval, paler than ground-color of wing, with dark central shade and black outline; reniform narrow at costal portion, broadening basally and bent outwards, similar to orbicular in color, with dark central shade, particularly well defined toward base of wing; space between spots dark velvety black, the most prominent feature of the maculation; a faint median shade perceptible below the reniform; t. p. line bent well outward below costa, slightly incurved below cell, composed of inner dentate black line and pale obsolescent shade, points of the teeth slightly pale-tipped; s. t. line indistinct, slightly ochreous, defined by some dark shading in subterminal area, parallel to outer margin and ending in an indistinct ochreous spot above anal angle; several ochreous costal dots before apex of wing and black terminal row of spots; fringes dusky with pale basal line. Secondaries smoky, paler at base, with discal dot and darker terminal line. Beneath, primaries smoky, secondaries whitish, heavily sprinkled with purplish and with prominent discal dot and black terminal line and traces of broken postmedial line. Expanse 28 mm.

HABITAT. Palmerlee, Ariz. (Biederman) 1 ♂. Type, Coll. Barnes.

The species resembles *agrotiformis* Grt. in general appearance, but the ciliate antennae would, according to Hampson, throw it into Section III of this genus; the black patch between orbicular and reniform renders it sufficiently distinct.

ERIOPYGA DUBIOSA sp. nov. (Pl. V, Fig. 1).

♀. Head, thorax and primaries a dull brown suffused with smoky, maculation not very distinct; basal line black; t. p. line black, single, waved, preceded by an ochreous line, representing the filling of an obsolete geminate line; t. p. line black, bent outwards below costa, then inwardly oblique, passing close to lower edge of reniform, dentate, bent outwards slightly above inner margin, followed by a pale ochreous shade-line; orbicular outlined in black, horizontally oval; reniform broad, not constricted, outlined in black, open towards apex of wing, filled with slightly paler shade than ground color of wing; claviform rather long, finger-like, black outlined; s. t. line represented by a series of ochreous dots preceded opposite the cell by two black dashes, of which the lower one extends across the subterminal space; in the interspace between veins 2 and 3 is also a faint small dash; fringes dark, checkered with ochreous. Secondaries smoky, hyaline towards base, with a distinctly dentate postmedian smoky line and dark terminal line. Beneath, primaries smoky, paler outwardly, secondaries hyaline, sprinkled with smoky, both wings crossed by a broad dentate smoky line, ending close to anal angle of secondaries, small discal dot on secondaries.

♂. Antennae serrate and fasciculate; primaries similar to those of ♀; secondaries white with almost no smoky sprinkling; terminal smoky line. Beneath whiter than ♀ without the prominent postmedial smoky line. Expanse ♂ 25 mm, ♀ 30 mm.

HABITAT. San Benito, Texas (Mch. 16-23, Sept. 8-15) (G. Dorner) 1 ♂, 1 ♀. Types, Coll. Barnes.

A second ♀ is before us, but too poor to make a cotype. The species bears considerable resemblance to *serrata* Sm. *(Trichopolia)*, placed by Hampson in this genus; but the structure of the ♂ antennae at once separates the two species. The abdomens are rubbed so we are unable to determine whether tufts are present or not; the receipt of fresh specimens may show that the species is more correctly placed in *Polia*.

MONIMA TENUIMACULA sp. nov. (Pl. V, Fig. 2).

Palpi, head, and pectus gray tinged with rosy, thorax olivaceous, abdomen ochreous, tinged with rosy in ♀; antennae in ♂ shortly bipectinate; primaries light fawn color, very even and smooth in appearance; t. a. line geminate, outer line more strongly marked than inner, brown, outwardly oblique to below cell, then straighter and angled inwardly on vein 1; orbicular wanting, reniform outlined in pale yellow, hour-glass shaped, very narrow, lower portion slightly filled with a gray-brown shade; median shade-line distinct, broad, brown, oblique to base of reniform, then subparallel to t. p. line and close to same; t. p. line geminate, inner line alone distinct, brown, crenulate, outcurved to beyond cell and then parallel to outer margin; a few black points on the veins and an indistinct connecting shade give the position of the outer line; s. t. line a row of distinct black spots, even, parallel to outer margin, most marked at costa; outer margin crenulate and slightly shaded with pinkish; secondaries, in ♂ smoky, shading towards ochreous at base, in ♀ uniform deep smoky; fringes rosy. Beneath ochreous, suffused with rosy with faint discal dots and subterminal lines. Expanse 34 mm.

HABITAT. Kerrville, Texas (Lacey) 1 ♂, 1 ♀. Types. Coll. Barnes.

We know of nothing resembling this very closely; the even smooth appearance without orbicular combined with the prominent shade line and narrow reniform give it a very characteristic appearance. It would evidently fall somewhere near *annulimacula* Sm.

Cucullia minor sp. nov. (Pl. V, Fig. 5).

Head and thorax gray mixed with blackish; tegulae not greatly produced, crossed by a curved black line, abdomen ochreous; primaries dark gray, transverse lines very indistinct, spots rather evident, inner margin shaded with blackish, veins slightly black; t. a. line geminate, smoky, waved below costa, with a very prominent outward angle in submedian fold extending below the orbicular, the line then receding to half-way between base of wing and apex of this angle, indistinct below anal vein; orbicular round, outlined in black, reniform broad at base, narrowing towards costa, outlined in black with slight central smoky lunule; t. p. line almost imperceptible except as an angular mark above inner margin, marked by some whitish dashes bordered with smoky on the veins, angled inwardly in submedian fold; s. t. line marked by smoky shading below costa and by a small dark patch above anal angle, which is distinctly cut by a pale waved line and extends along vein 2 to outer margin as a dark dash; slight terminal dark shade-lines opposite cell; broken dark terminal line most distinct towards anal angle; fringes dusky with pale basal line. Secondaries pure white with dark terminal line. Beneath, primaries smoky, secondaries hyaline white. Expanse 33 mm.

Habitat. Deming, N. M. (Sept. 1-7) 2 ♂'s. Types, Coll. Barnes.

The species resembles a *Copicucullia* but lacks the terminal claw to fore tibiae; the well defined spots along with the broken character of the dash above tornus are sufficiently characteristic to make the species readily recognizable.

Oncocnemis linda sp. nov. (Pl. V, Fig. 4).

Front deep black-brown, head and collar gray, latter intermingled with orange scaling and tipped with white; thorax pale whitish-gray; primaries white, scaled with golden-ochreous and heavily suffused terminally with smoky-brown; basal line black, waved; t. a. line arising from a small black patch on costa, black-brown, strongly outwardly oblique with outcurves in submedian fold and above inner margin; subbasal space with slight ochreous and smoky scaling; t. p. line with black patch on costa at origin, outcurved around cell, slightly incurved in submedian fold, dentate, brown; orbicular white, oval, edged with ochreous, reniform white, broad, indistinctly outlined in ochreous, with dusky central lunule; a diffuse brownish ochreous median shade arising from black patch on costa and passing between orbicular and reniform, angled outwardly on vein 1; beyond t. p. line a narrow band of ground color followed by heavy smoky shading extending to outer margin and slightly tinged with golden ochreous; the s. t. line is pale and defined by still deeper shading in the subterminal space which is wanting opposite cell, but very heavy at costa and above inner margin; a terminal dark lunulate line; fringes white, checkered brown and ochreous at base. Secondaries hyaline white with broad sharply defined outer smoky border and pale fringes; in ♀ basal portion slightly ochreous,

Beneath, both wings whitish, with broad smoky terminal border preceded by an incomplete subterminal line; fringes pale. Expanse 27 mm.

HABITAT. Loma Linda, S. Bernd. Co., Calif. (Pilate). 1 ♂, 1 ♀. Types, Coll. Barnes.

In general style of maculation the primaries resemble *regina* Sm. but the hyaline white secondaries with dark outer border are sufficiently characteristic of the present species to warrant a name.

GRAPTOLITHA (XYLINA) LACEYI sp. nov.

♂. Head and thorax pale gray deepening in color posteriorly, collar crossed by dark line near apex; primaries gray, maculation distinct and sharply defined; pale gray patch on costa at base extending downwards to black basal dash and crossed by a faint dark gray dentate subbasal line; t. a. line geminate, inner line gray, outer blackish, filled with pale gray, dentate below costa, with strong inward angles on median and anal veins and a distinct prominent W mark below cell, outwardly angled above inner margin; claviform outlined in black, filled with dark gray, resting on teeth of W mark; orbicular large, slightly oblique, constricted towards base giving figure of 8 appearance, outlined in black, filled with pale gray, basal portion filled with dark gray shade; reniform upright, outlined in black, filled with ground color, with a slight inward angle along median vein towards base of wing, this being tipped with white, and a corresponding angle towards apex of wing; a dark median shade crossing lower inner portion of reniform and angled inwardly in submedian fold; t. p. line geminate, rather indistinct. outer line gray, inner line blackish, prominently dentate opposite cell and on anal vein, contiguous with median shade below reniform; s. t. line dark gray, accentuated by a series of prominent arrow like dashes preceding it in subterminal area between inner margin and vein 6, angled outwardly below costa, then parallel to outer margin; terminal dark line; fringes concolorous. Secondaries smoky, paler basally, with discal dot. Beneath, primaries smoky, paler along costa and outer margin, with small discal spot; secondaries whitish, sprinkled with smoky with waved postmedian line and large discal spot. Expanse 43 mm.

HABITAT. Kerrville, Texas (Nov.) (Lacey). 1 ♂. Type, Coll. Barnes.

Belongs in the *cinerosa* group, but may be distinguished by its paler color and sharper, more contrasted markings with more prominently dentate t. a. line. We take pleasure in naming it after the collector from whom we have received so many new and interesting species; it has been figured in Vol. II, No. 1 of our Contributions.

PARASTICHTIS LIGNICOLORA ATRICLAVA subsp. nov.

Deeper ochreous and more uniform in coloration than the type form from the East, without the prominent contrasts between the pale maculation and the ground-color; the main distinction consists in the claviform being distinctly outlined in black-brown, whereas in the type form it is absent; the shape and position of the orbicular very variable, in general large, obliquely oval, with rather ragged edges; the course of the ordinary lines as in the type form. Expanse 45-49 mm.

HABITAT. Duncans, Vanc. Is., B. C. (Hanham). 4 ♂, 6 ♀. Types, Coll. Barnes.

·We consider this a Western race of the well known Eastern species; we have made three or four slides of the ♂ genitalia of each species and note the following slight distinctions; in the Western form the apex of the uncus is distinctly more spatulate than in *lignicolora;* the gnathium (scaphium) is much reduced, being short and semiquadrate, whereas in *lignicolora* it is long and pointed; the apical lobe of the valves or claspers is smaller and slightly differently shaped in the western form and there are several slighter points of detail which can be better seen than described. Smith's figure (Proc. U. S. Nat. Mus, XIII, Pl. XXXVII, Fig. 29) gives a general idea of the shape of the clasper and harpe. The species was figured in Vol. II, No. 1 of our Contributions.

We have before us another form from Vancouver Is. which shows distinctions in both coloration and ♂ genitalia sufficient in our opinion to warrant it being considered a good species. The ♂ claspers are much broader at the base than in either *lignicolora* or *atriclava;* the uncus apex is as in *atriclava,* but the gnathium much as in *lignicolora;* the lateral portions of the tegumen are broader and the apical portion of same, bearing the hair pencil, much less sharply pointed; we propose the following name:

PARASTICHTIS PURPURISSATA sp. nov.

Primaries deep purple-brown, washed in subterminal area, in ♂ slightly, in ♀ more strongly, with whitish, much as in *auranticolor* Grt. Markings as in *atriclava.* Expanse 45-48 mm.

HABITAT. Duncan, Vanc. Is., B. C. (Hanham) 3 ♂, 3 ♀. Types, Coll. Barnes.

The species is easily recognized by its deep purple brown color; it appears to be rarer than *atriclava,* of which we have numerous specimens besides the types; breeding will be necessary to prove the exact relation of these two forms; it was also figured in Vol. II, No. 1 of the Contributions.

113

PARASTICHTIS ATROSUFFUSA sp. nov. (Pl. III, Fig. 13).

♀. Bright reddish ochre, strongly suffused with deep purple-black, leaving the ground color apparent only at base of wing, around the reniform, orbicular and claviform, and on portions of the s. t. line. Markings much as in other members of the group; fine black streak from base to claviform; t. a. line outwardly oblique with prominent inward angle on vein 1; ordinary spots outlined in brown, filled with bright ochreous; orbicular irregularly oval, large, very oblique; reniform upright, figure-of-eight shaped, broadest at base which is shaded with blackish; claviform large, distinct; t. p. line crenulate, well outcurved around reniform; s. t. line preceded by blackish streaks, bright ochreous with prominent W mark. Secondaries smoky, deeper outwardly. Expanse 48 mm.

HABITAT. White Mts., Ariz.; Poncha Spgs., Colo. 2 ♀. Types, Coll. Barnes.

The specimen from Colo. has been in the collection a long time, together with two rather battered ♀ specimens from Durango, Colo. The receipt this year of a fresh, beautiful specimen from White Mts., Ariz., makes it evident that we are dealing with at least a good race, extending through Southern Colorado along the Arizona-New Mexico border. The heavy blue-black suffusion of primaries is very characteristic.

OXYCNEMIS ERRATICA sp. nov.

Head and thorax olivaceous brown sprinkled with white, with a faint black line across collar and similar ones along upper edges of patagia; metathoracic tufts darker; abdomen with dorsal tufts on basal segments; primaries olive-brown, paler basally and suffused with white terminally; subcostal vein marked in black from base to halfway across subbasal space; a black dash below cell connecting with lower edge of claviform; t. a. line black, single, upright, curved in slightly at costa and with inward angle below claviform; claviform large, semicircular, resting on t. a. line, joined to t. p. line by a faint black streak; orbicular and reniform large, round, subequal, outlined in black, scaled partially interiorly with white; t. p. line black, single, angled prominently outwardly below costa, then inclined inward to vein 2, touching lower portion of reniform in passing, forming between vein 2 and inner margin a strong even incurve shaded outwardly with white; subterminal space shaded with white, which more or less clearly defines the s. t. line; this line is very irregular, angled slightly outwardly below costa, almost touching outer margin at veins 3 and 4, receding at vein 2 to t. p. line and again broadening towards inner margin, leaving in terminal space a triangular patch of olive brown through which a black dash runs from t. p. line to outer margin along vein 5; fringes checkered ocherous and smoky. Secondaries white, with faint smoky terminal line and ends of veins tipped with same color. Beneath, white, primaries sprinkled with smoky and with dusky costal spot beyond cell, fringes checkered; secondaries immaculate. Expanse 29 mm.

HABITAT. Brownsville, Tex., San Benito, Tex. (Mch. 16-23) (Dorner) 1 ♂, 4 ♀. Types, Coll. Barnes.

The species is very close to our *grandimacula* from Arizona, differs in the subterminal white shading and lack of black dash on vein 4 from reniform outwardly; it may prove to be merely a geographical race of this species. The tufting on abdominal segments is not typical of the genus *Oxycnemis* as defined by Hampson, but we do not for the present care to erect a new genus based on this feature; the other structural details agree with *Oxycnemis*. Figured in Vol. II, No. 1 of our Contributions.

LEUCOCNEMIS SUBTILIS sp. nov. (Pl. V, Fig. 13).

Palpi, head, and base of collar deep chocolate brown, collar whitish sprinkled with olive, thorax and primaries rather pale gray-brown, sprinkled with white; basal half-line single, black; t. a. line single, black, bent slightly at costa, then rigidly outwardly oblique; claviform large, oval, outlined in black, either resting on t. a. line, connected with it by a short dash, or entirely separate; orbicular round, outlined in black, variable in size, usually large; reniform large, oval, not constricted; a faint median shade line passes between the two spots and is continued as a sinuate line to inner margin close to t. p. line; t. p. line black, single, outwardly oblique from costa to beyond reniform and often touching apex of same, curving around reniform and usually touching base, then sinuate to inner margin; s. t. line rather faint, indicated by white scaling, preceded by slight dark dashes which may be obsolete, slightly dentate and irregular; black dash from middle of reniform to s. t. line; veins terminally slightly outlined in black; a black terminal line; fringes slightly checkered. Secondaries white, with faint dotted subterminal line and apical dark shade extending downward along outer margin at times; terminal dark line not attaining anal angle; fringes white. Beneath primaries smoky; secondaries much as above. Expanse ♂ 25 mm., ♀ 27 mm.

HABITAT. San Benito, Texas (Mch. 16-23) Brownsville, Tex. (Dorner) 4 ♂, 5 ♀. Types, Coll. Barnes.

Very close to *sectilis* Sm. of which species we have two specimens before us. True *sectilis* is pale "ashen gray" in color, our species much darker, being gray-brown; in *sectilis* the dark dash from reniform to s. t. line is wanting and the subterminal area is almost uniform in coloration without s. t. line; the t. p. line in *sectilis* is slightly angled inwardly on vein 1, as may be seen in Smith's figure; in our species it is straighter. The relative positions of spots and lines are rather variable.

ALEPTINA INCA Dyar. (Pl. V, Fig. 15).

This species was described from two ♂'s from Arizona and 2 ♀'s from Texas; a pair of these are in Dr. Barnes' Collection. The receipt of two ♂'s recently from Texas makes it apparent that we have two distinct races of this species and we would therefore restrict the name *inca* to the Arizona form of which a long series is before us and which is characterized by the prominent yellow patch at base and more distinct maculation. The Texan form we characterize as follows:

A. INCA TEXANA subsp. nov. (Pl. V, Fig. 14).

Primaries paler than in the typical form, suffused with bluish-gray, maculation being more or less indistinct; the yellow patch at base is almost wanting, being represented by a pale ochreous suffusion; the subterminal and terminal areas are scarcely differentiated and the s. t. line is faint; the apical white shading is less prominent, being more suffused with the blue-gray ground color. Expanse 20 mm.

HABITAT. Brownsville, Texas (Mch. 8-15) (Dorner), Kerrville, Tex. (Lacey) (♀) 2 ♂, 1 ♀. Types, Coll. Barnes.

ATHETIS MINUSCULA sp. nov.

♀. Palpi outwardly black, tip of 2nd joint, 3rd joint, and inner side white; head and thorax light purplish-gray, primaries even deep purple-gray, all maculation indistinct; t. a. line indicated by a black dot at costa, another in the cell and three irregularly placed above inner margin; orbicular two black dots connected by a faint curved u-shaped mark, open towards costa; reniform more distinct, broad, partially outlined in smoky with some scattered whitish points at base and apex, it may or may not be filled with black; t. p. line indicated by two widely separated curved rows of parallel black dots arising from two larger dots on costa; a rather thick dark terminal line preceded by a faint ochreous hair line above tornus; fringes dark with pale basal line. Secondaries white, rather hyaline, with slight smoky shading most pronounced at apex and continued along outer margin not attaining anal angle; fringes white, tinged with smoky at apex. Beneath, primaries smoky with faint discal dot; secondaries white sprinkled along costa with smoky. Expanse 25 mm.

HABITAT. Brownsville, Texas (Dorner) 2 ♀. Type, Coll. Barnes.

The species in size and general coloration resembles our *mona* from S. Calif. differs however in the details of maculation; the cotype has the reniform rather heavily filled with black basally and towards costa, leaving a paler transverse central bar of ground color; in most N. Am. species of this genus this will be found to be a variable quantity; the species was figured in Vol. II. No. 1 of our Contributions.

CATABENA SAGITTATA sp. nov. (Pl. V, Figs. 7, 8).

♂. Head and thorax gray, latter with a prominent hood; primaries gray, rather strigate in appearance due to slight broken black marks on the veins; ordinary maculation obsolescent, slight oblique marks on costa indicate t. a. and t. p. lines; reniform very indistinct, indicated by a pale oval spot at end of cell, below reniform a darker shade and below this again a whitish suffused blotch between vein 2 and anal vein, tending to become drawn out into points toward anal angle; in the interspaces with their bases resting on outer margin a series of long black arrow-like marks edged and tipped with white, those at anal angle becoming merged in the whitish patch already mentioned, this being the most distinct portion of the maculation; fringes gray, cut with white. Secondaries white, rather hyaline, edged outwardly with smoky. Beneath, primaries largely smoky, paler above inner margin; secondaries white sprinkled with gray scales along costa and at apex and with dark terminal line, fringes white.

♀. Darker in color than the ♂, more blue-gray, secondaries with a somewhat broader smoky boarder, maculation similar. Expanse 27 mm.

HABITAT. Loma Linda, S. Berndo. Co., Calif. (Oct. 8-15) (Pilate). 3 ♂, 4 ♀. Types, Coll. Barnes.

The species is allied to *lincolata* Wlk., is however darker in color and differs in the details of the maculation.

PROXENUS MINDARA sp. nov. (Pl. V, Fig. 9.)

♂. Head, thorax and primaries fuscous-brown, paler than in *miranda* Grt. and not so glossy; veins in terminal area streaked with dark fuscous; t. a. line wanting; t. p. line represented by a series of dark points on the veins, followed by a paler shade line; orbicular a black dot, reniform incompletely outlined, small, represented by a lunate black mark edged towards base of wing with white and a white dash partially defined on both sides by black; faint terminal row of dark points; fringes concolorous, with pale dots opposite veins; secondaries whitish, with smoky discal dash and apical suffusion and faint dark terminal line; fringes pale, cut by darker line near base. Beneath, primaries deep smoky with slight ochreous sprinkling and dark terminal dotted line; secondaries white, heavily sprinkled in costal half with smoky and with dark discal dash and terminal broken line. Expanse 25 mm.

HABITAT. San Bernadino Co., Calif. (Oct. 8-15) (Pilate). 2 ♂. Type, Coll. Barnes.

The species is closely allied to *miranda* Grt., but paler and with more details of maculation. We know of no records of *miranda* from this region. In the cototype the reniform is very indistinct with only slight traces of white.

GORTYNA SERRATA forma OCHRIMACULA for. nov. (Pl. III, Fig. 11).

♂. Maculation as in typical *serrata* Grt. from Colo. but the spots are all deep yellow instead of white.

HABITAT. White Mts., Ariz. (Lusk) 1 ♂. Type, Coll. Barnes.

This may prove to be a geographical race when more material comes to hand; we refer it to *serrata* Grt. rather than to *repleta* Bird, recently described from Arizona, as the shade of the hind-wings agrees exactly with that of the former species.

OGDOCONTA SEXTA sp. nov. (Pl. V, Fig. 6).

Head, thorax and primaries deep olive-brown, sprinkled more or less heavily with white scales; t. a. line indistinct, whitish, outwardly oblique from costa to median vein, then inwardly oblique and subparallel to t. p. line; orbicular and reniform wanting; t. p. line white, bent strongly outward below costa, then inwardly oblique and rather rigid to beyond centre of inner margin, usually indistinct at costa, followed by slight pale shading in subterminal space and with a more or less distinct white shade proceeding from its costal angle to apex of wing; 4 pale dots on costa apically; s. t. line defined by some slight dark shading in subterminal space, indistinct, slightly crenulate, but generally rigid and subparallel to t. p. line; a terminal lunulate dark line; fringes concolorous. Secondaries smoky brown with paler fringes. Beneath, primaries smoky, dusted with whitish outwardly and with indistinct discal lunule; secondaries paler with broad subterminal dark line. Expanse 25 mm.

HABITAT. Brownsville, Tex., San Benito, Tex. (Mch.-June) (Dorner). 4 ♂'s, 7 ♀'s. Types, Coll. Barnes.

Closest to *altura* Barnes, but differs in the course of both t. a. and s. t. lines.

EUSTROTIA CNOSSIA Druce (Pl. V, Figs. 10, 11).

Pseudina cnossia Druce, Biol. Cent. Am. Het. 1, p. 311, Pl. 28, Fig. 28 (1889).

Eustrotia cnossia Hampson, Cat. Lep. Het. X, p. 602 (1910).

We have a long series of what is apparently this species from Brownsville and San Benito, Texas.

TARACHE DORNERI sp. nov. (Pl. IV, Figs. 8, 9).

Head and thorax white; entire basal half of primaries white, this color extending broadly along costa to near apex; the white portion of the wing is bordered outwardly by an olivaceous broad line, straight and parallel to outer margin from costa to just below vein 6, then turning at almost right angles and running inward parallel to costa below radius to just before inception of vein 11, again sharply angled and slightly inwardly oblique to center of inner margin; beyond this line the outer half of wing is deep purple with traces of a subterminal line showing above anal angle as a white waved line preceded by olivaceous scaling; about middle of costa a triangular purple patch shaded with olivaceous and almost connected with dark portion of wing; beyond this a small round purple reniform, situated on the olive line, with its upper edge touching the white costal portion, ringed with white and with scattered white central scales; a broken dark terminal line, preceded by a fine white line; basal portion of fringes purple, heavily suffused with white, outer portion pure white. Secondaries white, suffused with smoky apically and outwardly. Beneath, primaries deep smoky, with white fringes, secondaries white with an elongate smoky patch on costa near apex. Expanse 16 mm.

HABITAT. San Benito, Texas (Mch. 16-23) (G. Dorner). 3 ♂, 3 ♀. Types, Coll. Barnes.

The species is closely related to *cretata* G. & R., is however smaller, with the marginal border to white area much more sharply defined and angled. We name the species after Mr. G. Dorner, from whom we have received numerous interesting species from our southern border.

TARACHE CURVILINEA sp. nov. (Pl. V, Fig. 12).

♂. Head and thorax pale ochreous, latter with a purplish tinge; primaries, basal half to t. p. line whitish, irrorate with brown and purplish scaling, the latter strongest in the median area above inner margin; a faint slightly curved and inwardly oblique smoky t. a. line; t. p. line white, with strong outward angle below costa and then evenly and strongly incurved to beyond middle of inner margin; beyond t. p. line the wing is purplish brown, the arcuate portion of t. p. line being filled with a deep chocolate brown lunate patch, bordered outwardly slightly with purplish and above inner margin by an oblique white streak; a terminal series of dots; fringes dusky. Secondaries hyaline white, very slightly ochraceous at apex. Beneath primaries ochreous with darker smoky shade towards apex of wing; secondaries hyaline. Expanse, 23 mm.

HABITAT. Brownsville, Tex. (June 11). San Benito, Tex. (April 1-7) (G. Dorner) 2 ♂. Type, Coll. Barnes.

The species belongs in the *terminimaculata* group and is evidently closely related to *cratina* Druce (Biol. Cent. Am. Pl. 28, Fig. 26); it lacks however the dot representing the orbicular, and the basal area of primaries is much paler. Our specimens are rather imperfect but from the two combined a fairly accurate description can be drawn up.

HOMOCERYNEA gen. nov. (Type *H. cleoriformis* sp. nov.).

♂. Antennae serrate and fasciculate, ♀ antennae ciliate; palpi upturned, 2nd joint rather flattened laterally and heavily scaled, reaching vertex of head, fringed slightly with hair above, 3rd joint long, pointed, smoothly scaled; front smooth; thorax clothed with flat scales, prothorax without crest, metathorax

119

with rounded, knob-like crest of metallic scales; eyes moderate, tibiae unarmed; primaries with termen well rounded, veins Cu₁ and Cu₂ approached from well before angle of cell, vein M₂ from above angle, R₃ and R₄ stalked from apex of areole, R₂ from areole, R₁ free; secondaries with M₃ and Cu₁ connate from lower angle of cell, M₂ nearly fully developed from just below middle of discocellulars, S. C. joined to cell only at base of wing.

The genus is apparently closely related to *Cerynea* Wlk. (Hampson, Cat. Lep. Het. X, 209). The rounded tuft of metallic scales on metathorax, somewhat similar to, but smoother than in *Oxycnemis* is one of the chief points of distinction. The specimens before us have had the thorax rather crushed owing to being papered but as far as we can judge there are no prothoracic crests.

H. CLEORIFORMIS sp. nov. (Pl. IV, Figs. 10, 11).

Distinctly geometriform in appearance, resembling a small specimen of the genus *Cleora;* dull white, heavily clouded and powdered with smoky-brown; a triangular brown costal spot near base of wing, beyond this a larger, more diffuse patch, marking the inception of the t. a. line; from this an indistinct wavy brown shade crosses the wing, preceded in the cell by a slight brown dot; reniform an oval brown spot; beyond cell and touching t. p. line a minute dot, at times obsolete; t. p. line most distinct feature of the maculation, white, dentate, bordered by brown scaling, outcurved below costa, incurved slightly opposite cell, then irregular and subparallel to outer margin; subterminal space shaded with brown, most noticeable as a costal patch and again above inner margin; subterminal line defined slightly by a brown terminal patch between veins 3 and 5, otherwise obsolete; prominent row of dark brown terminal lunules; fringes white, checkered with brown. Secondaries very similar to primaries in general appearance, marbled and suffused with brown, with faint continuation of t. p. line of primaries and rather prominent brown patch above anal angle; small brown dot in cell; inner margin rather paler and crossed by several brown lines; fringes and terminal line as on primaries. Beneath, primaries suffused with smoky, slightly ochreous along costa with indistinct brown markings; secondaries paler, with traces of ante-median and post-median lines and discal dot; fringes of both wings as above. Expanse 22-25 mm.

HABITAT. Palmerlee, Ariz. (Biederman) 1 ♂, 3 ♀. Types, Coll. Barnes.

We have been unable to find anything at all corresponding to this species; it is possible it has been described as a geometer, but if so we have not succeeded in identifying it.

GEOMETRIDAE

It is with some hesitation that we venture to describe the following species, as it is quite within the bounds of possibility that they may prove to be Mexican or West Indian. We have however consulted Druce and Cuenée as fully as possibly without finding descriptions or figures that would suit any of these species; Mr. J. A. Grossbeck has also examined the larger proportion of these and pronounced them as unknown to him. We therefore consider it better to bestow names upon them since the types of the Mexican species are so scattered as to render comparison a very long and tedious task; our figures should render identification easy, and allow those who have better opportunities of examining the types than we have to draw their conclusions. Unless the species are misplaced generically in our present lists we believe there are at present no N. Am. names that we can apply to our species.

XANTHORHOE COLUMELLOIDES sp. nov.

♂. Palpi, head and thorax ochreous sprinkled with brown, antennae strongly bipectinate; primaries crossed by alternate white and brown scalloped bands; these in the basal third of wing are rather indistinct and inconspicuous, about five in number; the median area of wing is crossed by five brown scalloped lines, the first and fifth corresponding to the antemedial and postmedial of other species, the antemedial line is preceded by a broad white line, defining the outer boundary of the basal area, lines two and three are widely separated at costa but approaches and are more or less conjoined below cell, forming a series of two or three pale diamond-shaped patches above inner margin; the large costal patch contains a brown spot in the cell; lines 4 and 5 are close together, the intervening space filled with pale ochreous, line 5 very strongly scalloped and followed by a broad white line; beyond this are three brown scalloped parallel lines, the intervening space filled with pale ochreous; this area is bordered outwardly by a prominent, broad, white, scalloped, subterminal line; terminal area brown, sprinkled with white scales and with broken brown terminal line; fringes white, checkered with smoky. Secondaries whitish with lines of primaries faintly reproduced in basal 2/3, but strongly marked in subterminal area, the white subterminal line and a preceding smoky line being particularly noticeable; terminal area even smoky brown with darker terminal line, fringes checkered. Beneath, basal portion of costa of primaries ochreous with a few brown streaks representing inception of brown lines of upper side; basal portion of wing smoky, paler along inner margin; subterminally the lines of upper-side are repeated; base of secondaries white slightly sprinkled with brown scales;.a brown discal dot followed by a broad median band, defined by brown scalloped lines and shaded within with light brown; subterminal and terminal areas as above, but the brown much lighter, not smoky; fringes on both wings as on upper side. Expanse 26 mm.

HABITAT. White Mts., Arizona (Lusk). 14 ♂. Type, Coll. Barnes.

. This species is apparently the same as that figured by Druce (Pl. 55, Fig. 13) as the ♀ of *Larentia columella;* our specimens, all ♂'s' do not however correspond with the figure and description of *columella* ♂ and we imagine that two species are represented. The name *columella* must be retained for the ♂ with immaculate secondaries.

XANTHORHOE DENTILINEA sp. nov. (Pl. VII, Figs. 4, 6).

♂. Antennae bipectinate, palpi smoky brown, head and thorax pinkish, sprinkled with black, abdomen pale fuscous, shaded dorsally with pink, with double row of black dorsal spots and white line at posterior portion of segments; primaries, basal portion deep purple-brown bordered outwardly by a pale yellow subbasal line and containing faint traces of cross-lines, most marked on inner margin; subbasal line oblique from costa to radius, then almost perpendicular and slightly waved; space between subbasal and antemedial lines paler, shaded with pinkish; antemedial line geminate, pale ochreous filled with purple-black, strongly and evenly oblique from costa to submedian fold, then with strong inward angle on vein 1, approaching close to subbasal line; median space to postmedial line filled by a broad band of deep purple-brown; the central portion of this band is paler and contains a connected series of irregular lozenge-shaped patches of which that in cell is the largest and contains a faint discal spot; post-medial line pale ochreous, defined outwardly by smoky line, incurved opposite the cell, outcurved at vein 4 and again incurved slightly in submedian fold; beyond this are traces of several parallel lines crossing the wing, marked on the veins by black dots; subterminal space pale pinkish with slight dark patches on costa; subterminal line white, lunate, broken, preceded opposite cell and above anal angle by slight dusky shading; an oblique ochreous dash from apex of wing to subterminal line below which the whole terminal area is shaded with rather smoky brown; a fine dark terminal line; fringes, inner half purplish, outer pale rosy, slightly checkered with smoky-brown. Secondaries pale smoky with slight rosy tinge, postmedial and subterminal lines of primaries continued distinctly across wing; traces of other lines basally and subterminally. Beneath primaries smoky, outer half rosy; postmedial line of upper side distinctly outlined in smoky and a few faint white dots in the position of subterminal line; faint discal dot; secondaries rosy, heavily sprinkled with smoky; small discal dot and fine waved postmedial line; dark terminal hair-line to both wings; fringes as above.

♀. Same as ♂, but pale areas much more ochreous and without the pink tinge. Expanse 25 mm.

HABITAT. White Mts., Arizona (Lusk). 3 ♂, 4 ♀. Types, Coll. Barnes.

The species belongs in the *convallaria* group but should be easily distinguished by the strong inward angle of antemedial line on vein 1; there is considerably variation in the distinctness of the transverse lines, as in all members of this group.

SYNELYS VESTALIALIS sp. nov. (Pl. VIII, Fig. 4).

Palpi, legs, and collar pale ochreous; front blackish; thorax, abdomen and wings pure white, latter with only faint traces of black speckles; primaries and secondaries crossed by three very faint, narrow, minutely waved, equidistant, ochreous lines, placed parallel to each other antemedially, medially, and post-medially; beyond the 3rd line and close to it can be perceived slight traces of a subterminal ochreous shade on some specimens; no discal spot present; faint ochreous terminal line; fringes pale. Beneath as above, but only 3rd line visible; small black discal dot present on both wings. Expanse 21-23 mm.

HABITAT. Vineyard, Utah (June 24, July 8, 10) (T. Spalding) 1 ♂, 3 ♀. Types, Coll. Barnes.

Closely related to *quadrilineata* Pack., which, according to our specimens, should be in genus *Synelys*, the ♂'s lacking all posterior spurs; it differs from this species in the purer white coloration and the narrower transverse lines, which are not rigid but slightly waved. The coloration will also distinguish it from *enucleata* Gn. from which it further differs in lack of discal spots on upper side and in the 3rd line being not nearly so irregularly waved, nor excavate opposite the cell. Our ♂ is rather worn, but the ♀'s very perfect.

LEPTOMERIS BENITARIA sp. nov. (Pl. VIII, Figs. 13, 15).

Antennae in ♂ ciliate, in ♀ simple; palpi above smoky-brown, below ochreous, front and a transverse line posterior to antennae deep smoky-brown, remainder of head and thorax pale creamy; wings pale creamy with very sparse black speckles; primaries crossed by five waved ochreous lines, the first ante-medial, outwardly oblique from costa to middle of cell, then inwardly oblique to inner margin 1/3 from base, often very indistinct; the second beyond the cell, parallel to first line but more waved, with slight incurve in submedian fold; the 3rd line parallel to 2nd and well removed from same, fine, minutely dentate, 4th line subterminal, rather broad and diffuse, irregularly waved; 5th line in the nature of a terminal shade, broadening towards apex of wing; a fine black dot in cell and series of terminal black spots. Secondaries with four ochreous lines, the 1st waved, either crossing or just preceding a small black discal dot; other lines similar and corresponding to lines 3, 4 and 5 of primaries; terminal black dots and pale fringes. Beneath almost unicolorous white, silky, with faint traces of lines of upper side on primaries, and terminal black spots to both wings. Expanse 15 mm.

HABITAT. San Benito, Texas (Apr. 24-30, May 16-23); Brownsville, Tex. (Dorner). 5 ♂, 6 ♀. Types, Coll. Barnes.

Rather like a pale *Eois ossularia* Hbn. but the structure would throw it, according to Hulst's tables, into the genus *Leptomeris*.

XYSTROTA ROSEICOSTA sp. nov. (Pl. VIII, Figs. 7, 8).

Antennae of ♂ lengthily serrate and fasciculate, almost bipectinate, of ♀ faintly ciliate; palpi purplish-red; fore and middle legs purplish red on basal joints, shading into ochreous on tarsi; posterior legs white; head, thorax and abdomen white; primaries white, more or less heavily sprinkled with fuscous scales, maculation very variable in distinctness, at times almost immaculate, well-marked specimens with following lines, smoky-brown;—an antemedial line strongly outwardly oblique from costa to cell, then parallel to outer margin, a median line slightly less bent at costa and with faint incurve in submedian fold, a postmedial line subparallel to median line with a distinct incurve opposite cell, these three lines equidistant from one another on inner margin; a discal dot midway between lines 1 and 2; a faint broken marginal line preceding the white fringes. Secondaries of the same color as primaries, antemedian line absent, other lines present but not directly continuing those of primaries, appearing on spread specimens to arise rather nearer base of wing; median line evenly curved, preceded by discal dot which at times is contiguous; postmedial line distinctly scalloped in central portion. Beneath white, with the *basal portion of costa of primaries pinkish, much more distinct in* ♂; primaries with discal dot and faint indication of postmedial line of upper-side, formed by dots on the veins; indistinct terminal broken line on both wings. Expanse ♂ 19 mm; ♀ 21-25 mm.

HABITAT. San Benito, Texas (Mch.-May) (G. Dorner); Brownsville, Tex. (Nov. 21). 3 ♂, 9 ♀. Types, Coll. Barnes.

Hulst's generic definition calls for no median spurs in hind tibae of ♀; our ♀'s of *hepaticaria* Gn. the type species, possess however both pairs of spurs; the same is true of the ♀'s of the present species; the ♂ antennae of our species are not so lengthily pectinated as in the ♂'s of *hepaticaria* and the hind legs less aborted, the tibiae having a tuft of hair on inner side. In venation the two species agree, two accessory cells being present in both cases, 8, 9 and 10 of primaries on long stalk and 6 and 7 of secondaries shortly stalked. We prefer to place the species in this genus rather than create a new genus based on ♂ characteristics only.

SCELOLOPHIA? DORNERARIA sp. nov. (Pl. VIII, Fig. 9).

♀. Palpi ochreous; front and collar brown, slightly bronzed, with a white line joining the bases of the antennae; thorax and wings opalescent white, latter more or less heavily sprinkled with brown scales; costal margin of primaries broadly brown, narrowing towards apex; an antemedial brown line, bent at almost right angles below costa, a slightly waved median brown line, in course almost perpendicular to inner margin; a submarginal brown line, heaviest at costa, undulate, strongly outwardly oblique to vein 3, where it closely approaches outer margin, thence parallel to same to anal angle; beyond this line the terminal area has a prominent deep brown patch at anal angle, another below vein 6, and a faint spot at apex of wing; a minute discal spot, a prominent terminal black-brown line; fringes brownish. Secondaries with lines 1 and 2 of primaries continued across wings as fine wavy brown lines; faint discal spot; dark terminal line and slightly brown terminal shading at apex of wing; fringes silky brown. Beneath whitish, costa of primaries and margin of wings slightly brownish; maculation of upper-side more or less visible on primaries. Expanse 19 mm.

HABITAT. Brownsville, Texas (Mar. 11, July 11) (G. Dorner). 3 ♀. Type, Coll. Barnes.

Having no ♂ the species is only temporarily placed in the above genus, as the venation shows two accessory cells. It is evidently related to *Acidalia olmia* Druce. We take pleasure in naming this delicate little species after the collector, Mr. Geo. Dorner.

DEILINEA UNDULARIA sp. nov. (Pl. VIII, Figs. 5, 6).

Front pale ochreous; thorax, abdomen and wings pure white, latter sprinkled more or less heavily with smoky scales; primaries crossed by three very obscure pale ochreous narrow lines of which the second is often obsolescent, the first or antemedial is bent sharply below costa, and the third is slightly curved around cell and obscurely scalloped between the veins; secondaries crossed by two similar lines, the outer one being a continuation of line 3 of primaries and showing distinct scallops on well-marked specimens. Beneath as above but transverse lines are missing; a faint black dot in cell on both wings and a slightly ochreous costal margin to primaries. Expanse 25 mm.

HABITAT. White Mts., Arizona (Lusk). 3 ♂, 4 ♀. Types, Coll. Barnes.

Very similar to *variolaria* Gn. but distinguished by the scalloped postmedial line of both wings.

GENUS SCIAGRAPHIA Hulst.

We have recently worked over a good deal of material from Southern Texas, belonging to what we term the *heliothidata* group. We find in the first place that these species are incorrectly placed in *Sciagraphia* as defined by Hulst, as all possess a strong groove and hair-pencil on inner side of hind tibiae and fall in *Macaria* as used by Hulst; we are further of the opinion that the synonymy as given in Dyar's list is not correct. What the true *heliothidata* Gn., which was described from Haiti, is, we do not know; however as Guenée's description fits in fairly well with a N. American species named *heliothi-*

data for us by Mr. Grossbeck we see no reason for changing the name at present; we figure this species (Pl. VI, Figs. 1-3); it may be recognized by the semihyaline basal 2/3 of wings and broad dark border both above and below, extending to margin of wings and tinged with yellow beneath; dark spots are usually present in this border above and below vein 4 on primaries just beyond postmedial line. *Ocellinata* Gn. appears to us to be a very good species and not a synonym of *heliothidata;* in fact it is the common Eastern species which apparently goes under the name of *punctolineata* Pack. according to our collection, which has been arranged by Mr. Grossbeck. Guenée's description applies excellently to this species and both Zeller (Verb. Z. B. Ges. Wien. XXII, 486, 1872) and Packard refer to it under Guenée's name, Packard's figure (Mon. Geom. Pl. X, f. 11) being an excellent representation; we figure the species (Pl. VI, Figs. 4-6); in general it may be recognized by the large discal spots with pale center, as the name implies.

Punctolineata Pack. proves to be the same species as that described later by Hulst under the name *Macaria simulata;* we have received a photograph of Packard's type through the kindness of Mr. C. Henshaw and Mr. J. Grossbeck, who has seen the type, confirms us in this opinion. The figure in the Monograph (Pl. X, Fig. 12) gives a very erroneous impression.

A third species before us is so close to *parcata* Grossb., described from Arizona, that we cannot point to any constant difference; in general it is rather paler and grayer in coloration; we figure this species (Pl. VI, Figs. 10-12). Several other species, of most of which we have long series, are apparently unnamed; we venture to describe them below.

MACARIA FLAVITERMINATA sp. nov (Pl. VI, Figs. 7-9).

Palpi and front gray slightly tinged with ochre; behind the orbits a small ochreous patch on collar; thorax and abdomen ochreous-gray sprinkled with black; wings hyaline, strongly dusted over with pale ochreous and sparsely sprinkled with blackish; this ochreous dusting becomes stronger beyond postmedial line and forms a broad grayish-ochreous terminal band to both wings; primaries crossed by three deep brown lines, usually more or less obsolete and appearing as if formed by an overlaying of brown scales on a pale orange ground; antemedial line bent below costa, then straight to inner margin; median line usually only indicated by an oblique dash or spot at costa, when present slightly irregular and crossing an obscure discal spot; postmedial line perpendicular to inner margin, bent inwards gradually towards costa from vein 6, heavily marked at costa and often between veins 3 and 4; beyond this line between veins 3 and 4 is usually a diffuse dark shade or spot which extends at times to inner margin but never above vein 4; a terminal row of minute black dots; fringes concolorous with terminal area. Secondaries with the lines usually obsolete and the dark marginal shade merging gradually into the paler basal color; at times lines 2 and 3 of primaries may be continued across secondaries as wavy lines semiparallel to outer margin; discal spot black, distinct; terminal row of dark spots; fringes as on primaries. Beneath as above but the contrast between inner and terminal area more marked; terminal area yellower, inner area paler; lines, except postmedian, mostly obsolete, dark shading in terminal band below vein 4 usually more distinct than on upper-side and often continued over on to secondaries; discal spots distinct on both wings. Expanse 20-23 mm.

HABITAT. San Benito, Texas (March-May), Brownsville, Tex. (G. Dorner). 6 ♂, 6 ♀. Types, Coll. Barnes.

A variable species but readily distinguished by the broad orange-yellow terminal band of underside; our figures represent a very strongly marked ♂, a typical ♀, and the underside of a ♀ with less dark shading in terminal area than usual, the spot alone being represented.

MACARIA SUBTERMINATA sp. nov. (Pl. VI, Figs. 13-15).

Antennae of ♂ serrate and fasciculate; palpi, head and thorax gray; ground color of wings whitish suffused with purplish-gray, especially heavily in outer third, forming a more or less distinct terminal broad band to wings beyond postmedial line; in some cases, notably ♀'s, this suffusion is reddish purple and the outer band very prominent; slight sprinkling of black scales; primaries crossed by three lines of which the inner two are poorly defined and at times obsolescent; antemedial line ochreous-brown, bent below costa, strongest at costa; median line waved, usually confined to a strong, very oblique costal dash and slight spots on median vein and inner margin; postmedial line pale ochreous, rigid, slightly bent inward at costa, preceded by a narrow brown shade line, thickest at costa; in the darker specimens the terminal area beyond postmedial line is uniformly dark, in paler specimens the terminal area is slightly lighter than subterminal, the subterminal line being represented by slight dark shading; all specimens possess a dark suffused patch just beyond postmedial line on vein 4; terminal row of black dots; fringes concolorous with terminal area. Secondaries much as primaries; median line of primaries continued across wing, more or less obsolete, with rather distinct spot at termination on inner margin; postmedial line distinct, as on primaries, separating pale basal area from darker terminal border; in pale specimens the subterminal area is deeper in color and is suffused with dusky shading, narrowing towards anal angle; small discal dot present. Beneath both wings with basal 2/3 whitish, slightly sprinkled with brown, with distinct discal dots; costa, apex and terminal area of primaries shaded with ochreous; a strong blackish subterminal shade, following course of postmedian line of upper side, narrow or wanting at costa, broadening below vein 6 and occupying whole subterminal area to inner margin; this shading is continued across secondaries, narrowing towards anal angle, mixed with ochreous, the line of demarcation between subterminal and terminal areas on secondaries very sharp with prominent tooth on vein 5; terminal area similar to basal; terminal row of black dots to wings; fringes ochreous. Expanse 21-25 mm.

HABITAT. San Benito, Texas (Mar.-Apr.); Brownsville, Tex. (June 11, Oct. 20) (G. Dorner). 6 ♂, 3 ♀. Types, Coll. Barnes.

The pale rigid postmedial line of upper side and the strong distinction between the dark subterminal and pale terminal areas of secondaries should serve to distinguish this species. Our figures show the type ♂ and ♀, the latter being one of the reddish-purple specimens, and a ♀ with typical underside.

MACARIA STIPULARIA sp. nov. (Pl. VI, Fig. 17).

♂. Head, thorax and abdomen gray; wings hyaline, whitish, heavily overlaid with gray scaling, heaviest on subterminal area of primaries and beyond postmedial line on secondaries; primaries crossed by the three usual lines; antemedial line dark gray, angled at right angles below costa; median line very indistinct, slightly waved; a small dark discal dot; postmedial line dark gray, gently bent outwards opposite cell, incurved in submedian fold, bordered outwardly by a faint pale shade-line; remainder of subterminal space deep gray, the darkest portion of the wing, with still darker diffuse shade patch between veins 3 and 4 and slight dark point on its outer margin between veins 6 and 7; terminal space lighter with terminal row of blackish dots. Secondaries with postmedial line of primaries continued, slightly waved; the whole space beyond it deep gray, slightly mottled with whitish; a blackish discal dot and faint traces of line preceding same; dark terminal dots. Beneath much as above, slight ochreous shading terminally, lines fainter, discal spots distinct; subterminal dark shade of primaries distinct towards inner margin. Expanse 20 mm.

HABITAT. Brownsville, Texas (Mch. 1-7) (G. Dorner) 1 ♂. Type, Coll. Barnes.

The species resembles *parcata* Grossb. in general coloration, but can be readily distinguished by the course of the postmedial line; in *parcata* this is sharply angled below costa and more or less punctiform; in our new species the line is gently bent outwards below costa and is not punctiform.

MACARIA ATRIMACULARIA sp. nov. (Pl. VI, Figs. 16, 18).

♂. Head and collar yellowish; thorax and abdomen grayish ochreous, former with slight purplish tinge; wings pale ochreous, heavily sprinkled with light brown and crossed by lines of the same color; antemedial line of primaries angled at right angles below costa and then perpendicular to inner margin;

median line in general parallel to preceding line; postmedial line outwardly oblique to vein 6 then parallel to median line, forming a sharp angle rather greater than a right angle on vein 6; all the lines contain more or less prominent dark dots on the veins; beyond the postmedian line a prominent black round patch between veins 3 and 4, and two smaller similar spots below costa; traces of a brownish dark shade line between the lowest dark patch and anal angle; very faint terminal dark dots; fringes concolorous. Secondaries strongly angled at vein 4; median and postmedial lines of primaries continued, former rather rigidly oblique, latter minutely waved; faint discal dot; a rather broad subterminal line, bent at vein 4; faint dark broken terminal line. Beneath paler and more yellowish than above, markings similar but rather more distinct; black subterminal spots of primaries hardly visible.

♀. Similar to ♂, but subterminal dark blotches almost lacking, the area beyond postmedial line being suffused with brownish, except apically. Expanse ♂ 21 mm., ♀ 23 mm.

HABITAT. Brownsville, Tex. (Mch. 11) (G. Dorner) 2 ♂, 2 ♀. Types, Coll. Barnes.

Belongs close to *punctolineata* Pack. but the lines are finer, sharper and more angled below costa and the prominent dark subterminal area of underside is lacking. In color it approaches *aequiferaria* Wlk. The three black subterminal spots of the ♂ are the most prominent features of the maculation.

DIASTICTIS GROSSBECKI sp. nov. (Pl. VII, Figs. 10, 12).

Head, thorax and primaries pale bluish-gray, latter more or less suffused with smoky in basal and subterminal areas and slightly sprinkled with smoky scales all over; antemedial line diffuse, smoky, bent below costa, slightly waved; postmedial line smoky, rather straight, slightly bent inwards in submedian fold, followed by more or less diffuse smoky shading, which tends to form dark patches beyond cell and above inner margin; subterminal line parallel to postmedial line, usually marked in costal half with four or five prominent black dots, which occasionally are obsolescent; median and terminal areas even blue-gray; small dark discal dot; fringes concolorous. Secondaries rather paler than primaries, at times with slight ochreous tinge, sprinkled with purplish atoms; with an evenly curved smoky postmedian line, parallel to outer margin, and faint discal dot; occasionally the terminal area is shaded with fuscous. Beneath primaries smoky, shaded with yellowish apically and with straight purplish postmedial line; secondaries ochreous, heavily sprinkled with purplish; line of primaries continuous across wing. Expanse ♂ 20 mm., ♀ 24 mm.

HABITAT. Brownsville, Texas (March); San Benito, Texas (March-April) (G. Dorner). 10 ♂, 9 ♀. Types, Coll. Barnes.

The species is related to *pallipennata* B. & McD. in type of markings; the costo-apical black dots and blue-gray coloration are characteristic; considerable variation is shown in the subterminal shading. We take pleasure in dedicating this species to Mr. J. A. Grossbeck, the well known student of Geometridae.

EUEMERA IMMACULATA sp. nov. (Pl. VIII, Fig. 2).

♂. Head, thorax and abdomen gray; primaries pale smoky-brown, more ochreous along costa and outer margin, which are thickly sprinkled with smoky atoms; faintest trace of a smoky postmedian line bent outwards around cell, not attaining inner margin; fringes gray; secondaries pale smoky, veins and outer margin slightly ochreous; fringes gray. Beneath primaries deep smoky brown, costa and apex gray, sprinkled with brown; secondaries even gray with heavy brown sprinkling and distinct discal dot. Expanse 29. mm.

HABITAT. Loma Linda, S. Calif. (July 8-15) (G. R. Pilate) 1 ♂. Type, Coll. Barnes.

The species resembles *simularia* Tayl. in general color but is paler, and the lack of maculation on both upper and under sides renders it easily distinguishable.

SELIDOSEMA NIGRICARIA sp. nov. (Pl. VII, Fig. 11).

♀. Primaries blackish, sprinkled, especially in median area, with white scales; antemedian line narrow, deep black, bent below costa and inwardly oblique to inner margin; median line broader, black, parallel to preceding line, beyond it a narrow discal dash; postmedian line slightly sinuate, scalloped between the veins, especially prominently between veins 1 and 2; subterminal line represented by white, diffuse, slightly scalloped markings preceded by slight black shading more prominent in central portion; terminal lunate black line; fringes deep gray. Secondaries with markings of primaries repeated except that antemedian line is wanting; subterminal line at times shows distinctly as a black line, bordered on both sides by white, but usually the white predominates and the black line cannot be traced; median area paler, due to sprinkling of white scales, with small discal dot. Beneath pale, silky, immaculate, sprinkled more or less strongly with smoky; small discal dots. Expanse 32-36 mm.

HABITAT. Palmerlee, Ariz. (C. Biederman) 1 ♂, 5 ♀. Types, Coll. Barnes.

The species resembles *gnophosarium* Gn. but can be most readily separated by the fact that the discal dots are punctiform and do not form ringlets. Our ♂ is in poor condition but possesses the hind tibiae which are without hair pencils so the generic reference is fairly certain.

SELIDOSEMA PURPURARIA sp. nov. (Pl. VII, Fig. 5).

♂. Head, thorax and abdomen suffused purplish gray, latter paler at base with slight black banding at rear of segments; primaries purple-brown, shaded with pale reddish-brown along inner margin and subterminally; antemedian line black, narrow, strongly excurved from costa to origin of vein 2, then inwardly oblique to inner margin; it is preceded by a broad blackish parallel diffuse shade line, attaining inner margin close to base of wing; median shade broad, black, diffuse, in general parallel to antemedian line, crossing a fairly large oval discal spot which shows faintly a pale central streak; postmedian line narrow, black, rather indistinct at costa and slightly emphasized on the veins, outwardly rounded from costa to origin of vein 3, then inwardly oblique to inner margin, with slight inward bend in fold, followed by a dusky shade, broadest at costa, and spreading towards outer margin in the interspaces of veins 4-6; subterminal line evenly scalloped, parallel to outer margin, preceded by pale purple-gray shading; terminal area darker. Secondaries with basal half to postmedian line whitish, heavily sprinkled with black; median band of primaries continued faintly across wing, most distinct on inner margin, which is purplish; postmedian line black, indistinct towards costa, very slightly scalloped, bent downwards towards anal angle from vein 2; large elongate discal spot; area beyond postmedian line purplish-gray, shaded, with subterminal line visible as a faint paler waved line; terminal black shading. Beneath whitish, silky, immaculate, with slight smoky sprinkling at costa of primaries. Expanse 41 mm.

HABITAT. White Mts., Ariz. (E. Lusk). 2 ♂. Type, Coll. Barnes.

This species had been doubtfully determined for us by Mr. Grossbeck from a single rather worn ♂ as *noctiluca* Druce (Biol. Cent. Am. Pl. 48, Fig. 7); the receipt of a second ♂ in excellent condition convinces us that we have at least a well marked race with distinct subterminal markings.

CLEORA ATRISTRIGARIA sp. nov. (Pl. VII, Figs. 1, 3).

Head and thorax whitish, collar tipped with black; abdomen ochreous, black at base followed by a broad whitish ring; wings mottled olivaceous, shaded with creamy along costa and reddish in subterminal space; lines indistinct at costa, otherwise very heavy and strongly marked; antemedian line black, outcurved to median vein then strongly inwardly oblique to inner margin, continued across secondaries by a black patch at base of wing; basal space shaded in lower portion with reddish next to antemedian line; median shade indistinct, narrow, rather undulate; discal spot very small and faint; median line rather outwardly oblique to vein 6, then strongly oblique inwardly to vein 2, with slight incurve in submedian fold and perpendicular from vein 1 to inner margin; subterminal area shaded with reddish, narrowing and fading out towards costa; from the cell a diffuse dark subtriangular shade extends outwards to margin of wing below apex; interspaces of veins 4-7 with 3 black dashes, the uppermost extending from outer margin just across subterminal line, the middle one extending 1/2 way across subterminal space and the lowest one attaining postmedian line; subterminal line whitish, scalloped, parallel to outer margin; terminal scalloped hair-line; fringes slightly checkered. Secondaries like primaries in coloration; with black postmedian line, not continuous with that of primaries, but arising considerably more outwardly, bent inwards below vein 7 and downwards towards anal angle below vein 1; a strong reddish shade follows this line; subterminal line whitish, scalloped, indistinct; small discal dot. Beneath primaries in ♂ largely smoky, with ochreous costa, apical patch and subterminal line and small dark discal spot, in ♀ basal 2/3 pale ochreous with diffuse broad smoky terminal band, leaving an ochreous apical spot free. Secondaries pale ochreous sprinkled with smoky, with terminal smoky band. Expanse ♂ 25 mm., ♀ 24-28 mm.

HABITAT. Brownsville, Texas; San Benito, Texas (May 16-23) (G. Dorner). 2 ♂, 2 ♀. Types, Coll. Barnes.

The species is allied to *pampinaria Gn.* but is much more contrastingly and heavily marked; the reddish subterminal shading and interspaceal black streaks are characteristic.

THERINA FLAVILINEARIA sp. nov. (Pl. VIII, Figs. 1, 3).

Wings in ♂ very slightly angled at vein 4, in ♀ more so; pale to dark ochreous, heavily sprinkled with smoky scales; antemedian line bent below costa, then almost perpendicular to inner margin, dark smoky brown, with prominent orange shading preceding it; postmedian line almost rigid, with slight bend at vein 4, parallel to outer margin, dark, followed by an orange shade; faint discal dot. Secondaries with oblique dark median line followed by orange shading, indistinct at costa. Beneath unicolorous pale ochreous with lines of upper side showing faintly through. Expanse ♂ 31 mm., ♀ 31-36 mm.

HABITAT. Palmerlee, Ariz. (C. Biederman), Redington, Ariz., White Mts., Ariz. 8 ♂, 3 ♀. Types, Coll. Barnes.

Allied to *cavillaria* Hulst but distinguished by the orange shade of the transverse lines and the deeper ground color. The relative positions of the lines in regard to each other are variable.

METANEMA HIRSUTARIA sp. nov.

♂. Front and palpi deep brown; thorax clothed with long rough hair, varying from reddish-ochre to purple brown; primaries varying in color similar to thorax, falcate and strongly angulate at vein 4, sprinkled with fuscous scales, with indistinct smoky evenly bent t. a. line and pale ochreous sinuate t. p. line, not bent at costa. Secondaries angled at vein 4, paler than primaries with more or less distinct ochreous postmedian line, preceded, especially towards inner margin, by smoky shading; small discal dot present on both wings. Beneath pale rosy ochreous with sprinkling of fuscous scales and small discal dot on both wings. Expanse 37 mm.

HABITAT. San Diego, Calif. 8 ♂. Type, Coll. Barnes.

This species seems best placed in *Metanema* although the rough hairy thoracic clothing is hardly typical of the genus; the venation corresponds to that of *quercivoraria* Gn. The species is variable in both coloration and position and intensity of the transverse lines; it was figured in Vol. II, No. 1 of our Contributions.

COSSIDAE

GIVIRA LUCRETIA sp. nov. (Pl. IV, Fig. 12).

Head and thorax gray, patagia sprinkled with whitish; primaries milky white shaded with deep brown; costa from base to apex with a series of small brown dots, cell with a few faint brown striations, otherwise immaculate; discocellulars whitish; beyond the cell between veins 2-6 a broad smoky-brown shade cut into a series of semiquadrate spots by the pale veins; at the inception of vein 2 this shade bends towards inner margin forming similar spots between the anal veins and becoming lost in a diffuse brown shade extending along inner margin from base of wing to anal angle, leaving the extreme margin at base of wing contrasting white; this shade is slightly broken by slight white striations; terminal portion of wing whitish with faint brown striations; ends of veins bordered on each side by dark dots, fringes pale, checkered with brown. Secondaries pale smoky with darker reticulations and intravenular dots as on primaries. Beneath primaries smoky, paler along inner margin, secondaries whitish, both striate with smoky brown especially terminally. Expanse 27 mm.

HABITAT. San Benito, Texas (Mch. 16-23, May 8-15, June 16-23) (G. Dorner). 3 ♂, 1 ♀. Types, Coll. Barnes.

The species is allied to *cornelia* N. & D., differs in the much greater prominence of the dark band beyond cell and the more prominent striations of secondaries.

NYMPHULINAE

Argyractis? confusalis sp. nov. (Pl. VIII, Fig. 11).

Primaries light brown, costa shaded with dark brown; an indistinct whitish subbasal line, rather broad and slightly oblique; a white median line, slightly dentate below costa and on median vein and then oblique to middle of inner margin, preceded along inner margin by some dark brown shading and followed for more or less its whole length by similar dark shade; a broad white oblique subterminal line from costa 1/4 from apex to vein 3 near outer margin, then abruptly and strongly incurved below submedian fold and bent outwards again to anal angle; beyond the cell between median and subterminal lines some obscure black markings which at times are centered with white and have appearance of two white spots separated by a dark streak; subterminal space sprinkled with dark brown; from costa near apex a white streak descends to vein 3, broadest at costa, nearly touching angle of subterminal line; bordered outwardly by fine dark line; dark terminal line; fringes smoky checkered with white. Secondaries white, a diffuse smoky subbasal patch in cell and another below middle of inner margin; an indistinct median smoky line, forming a prominent bar at end of cell then bent inwards towards spot near inner margin, turning again on vein 1 at right angles and attaining inner margin near anal angle; beyond median line costal 1/2 of wing is sprinkled with smoky scales; a terminal row of 6 small black spots between veins 2 and 6, separated by metallic scales, a terminal yellowish line broadest at costa; fringes white lengthening towards anal angle. Beneath much as above but paler with markings semiobsolete. Expanse 19 mm.

Habitat. White Mts., Ariz. 6 ♂. Type, Coll. Barnes.

Vein 4 of secondaries is absent, the palpi are upturned with 3rd joint acuminate; the species would thus fall according to Hampson in the genus *Argyractis* (Tr. Ent. Soc. Lond. 1897, p. 135) if it were not for the fact that vein 10 of primaries is separate from 8 and 9, arising from a point with same in all our specimens. This would in fact throw the species among the Pyraustids but as we know of no genus in this subfamily with vein 4 absent to which it could possibly belong, and as further the general appearance is so distinctly *Nymphulid,* we prefer to retain it in this association.

CHRYSAUGINAE

Genus GALASA Wlk.

If Hampson's figure and description of the venation of this genus (Proc. Zool. Soc. Lond. 1897, p. 674-5) be correct we have grave doubts as to whether the name *Galasa rubidana* Wlk. will apply to our N. American species. The species *rubidana* was described from Jamaica, and *Cordylopeza nigrinodis* Zell. (1873, Verb. Z. B. Ges. Wien. p. 206), described from Massachusetts, is listed by both Hampson and Dyar (Bull. 52 U. S. N. M. 401) as a synonym. We have examined the venation of a series of both ♂'s and ♀'s of our N. Am. form and find that what may be called the normal venation of primaries is as follows, just as stated by Ragonot (Ann. Soc. Ent. Fr. 1891, p. 508): ♂ ; veins 2 and 3 on very long stalk, 4, 5 from a point at angle of cell, 6 from upper angle, 7 and 8 stalked, 9 on slight stalk or connate with 7 and 8, *10 and 11 absent.* ♀ ; differs from ♂ in that 7, 8 and 9 are on longer stalk and 10 is usually present from just before angel of cell (in some specimens 10 is merely rudimentary and may even be absent). This would practically correspond with Zeller's figure of the venation of *Cordylopeza* and we consider it the safer course to refer to our common N. Am. species as *Cordylopeza nigrinodis* Zell. until our colleagues in the British Museum can verify Hampson's figure by an examination of the type specimen of *rubidana*. Certainly Walker's description is unrecognizable.

We have before us a closely related species from Arizona which we characterize as follows:

CORDYLOPEZA NIGRIPUNCTALIS sp. nov. (Pl. IX, Figs. 5, 6).

♂. Palpi, head and tegulae ochreous, suffused with red, abdomen ochreous, banded with smoky; thorax and primaries red, latter heavily sprinkled with ochreous scales; costal margin twice indented, this portion of the wing being more or less suffused with black scaling; t. a. line arising from first sinus, pale ochreous, heaviest at costa, slightly irregular in outline but not distinctly dentate, nearly straight in general course; t. p. line arising from second costal sinus, pale ochreous, strongly bent outward below costa and then parallel to outer margin; veins beyond it marked with black dots; veins of median area also more or less marked with black; a terminal row of prominent black dots, fringes purplish; secondaries pale ochreous, with dark, slightly broken terminal line extending backward along vein 2; fringes tinged with purplish apically. Beneath, primaries, costal 1/2 of wing suffused with red; lower half pale ochreous, costa shaded with black with two prominent ochreous spots in indentations; secondaries whitish, costa reddish, shaded with black, with a prominent ochreous dash near apex of wing; terminal broken dark line to both wings.

♀. Primaries excavated but once in long sweeping curve; pale cross lines indistinct, the black shading however being often more prominent and continuous than in ♂ ; a slight dark discal dot just beyond position of t. a. line; more or less diffuse ochreous shading around costal sinus; terminal dark dots not so prominent as in ♂, often wanting; secondaries smoky. Beneath much as in ♂ but coloration generally deeper. Expanse 18-20 mm.

HABITAT. Palmerlee, Ariz., Redington, Ariz. 3 ♂, 3 ♀. Types, Coll. Barnes.

The lack of the bright orange basal area should distinguish this species from *rubidana* Wlk. The species varies considerably in the distinctness of the maculation, some specimens showing scarcely a trace of same.

NEGALASA Gen. nov. (Type *N. fumalis* sp. nov.).

Palpi porrect, slightly downcurved at apex, about twice length of head; antennae annulate, in ♂ faintly ciliate; mid tibiae and hind tarsi tufted with hair; primaries in ♂ with costal margin twice excavated, with slight glandular swelling at base; 10 veined, 2, 3, 4, 5 from almost a point at lower angle of cell,

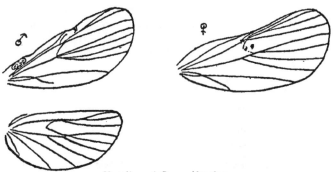

Venation of Genus *Negalasa*

at times 2 and 3, or 4 and 5 very slightly stalked, 6 from upper angle, 7 and 8 stalked, 9 connate with 7 and 8, from upper angle of cell, 10 and 11 absent; ♀ with single costal excavation, 11 veined, 2 and 3 separate at base and semi-parallel; 4 and 5 from a point or slightly stalked; 6 from upper angle, 7 and 8 stalked, 9 very shortly stalked with 8, 10 from cell, 11 absent. Secondaries in both sexes with 2 from well before lower angle of cell, 3 from angle, 4 and 5 slightly stalked, 6 from upper angle, 7 and 8 stalked.

Differs from *Cordylopeza* in that 2 and 3 of primaries are not on a long stalk but semi-parallel.

N. FUMALIS sp. nov. (Pl. IX, Figs. 3, 4).

♂. Palpi, head and thorax pale ochraceous; tufts on legs smoky brown, primaries, pale to deep ochraceous, slightly sprinkled with smoky scales and shaded with smoky along costa, crossed by two faint waved ochreous lines; the t. a. line is upright, bordered inwardly by a black shade, usually more prominent than the line itself; t. p. line strongly excurved, from 2nd costal sinus, then subparallel to outer margin, with prominent black border; faint trace of dark terminal line; fringes ochreous, smoky outwardly and apically; secondaries pale ochreous, slightly smoky, especially along vein 2. Beneath, primaries deep ochreous, secondaries pale, costa of both wings shaded with black; traces of postmedian smoky line marked on costa with ochreous spot.

♀. Primaries much deeper in color than in ♂, smoky brown, crossed by two indistinct black lines, corresponding to the black shade-lines of ♂; secondaries smoky. Expanse ♂ 15 mm., ♀ 18 mm.

HABITAT. San Benito, Texas (March, May, July), Brownsville, Texas (Oct. Nov.). 5 ♂, 4 ♀. Types, Coll. Barnes.

NEGALASA RUBRALIS sp. nov. (Pl. IX, Figs. 1, 2).

Primaries reddish with slight ochraceous tinge, almost immaculate; two very faint pale cross-lines are just discernable, in course much as in the preceding species and bordered with slight purplish scaling; terminal line and fringes purplish. Secondaries pale ochreous with faint purplish terminal shade-line, in ♀ somewhat smoky. Beneath primaries largely rosy, deeper along costa and with the indentations suffused with ochreous; secondaries pale ochreous, costal border rosy with ochreous postmedian dash preceded by darker shade. Expanse ♂ 16 mm., ♀ 18 mm.

HABITAT. Redington, Ariz. (Chrisman). 6 ♂, 1 ♀. Types, Coll. Barnes.

PENTHESILEA SACCULALIS Rag. (Pl. IX, Figs. 7, 8).

P. sacculalis Ragonot, Ann. Soc. Ent. Fr. 1891, p. 493.

This species, described from a single ♂, presumably from North America, has been omitted from our lists; we have however recently received several specimens from Southern Pines, N. C.; San Benito, Texas; and Babaquivera Mts., Ariz., which prove the species to be North American and to have a wide-spread distribution.

Chalinitis albistrigalis sp. nov. (Pl. VIII, Fig. 14).

Head and thorax whitish; primaries pale green, crossed by an oblique rigid white line from middle of costa to inner margin 1/3 from base; beyond this line costa is narrowly white to apex of wing; very faint traces of a white subterminal line; fringes concolorous. Secondaries white, slightly smoky. Beneath whitish, primaries and legs very faintly tinged with rosy. Expanse 20-23 mm.

Habitat. San Benito, Texas (Mch. 16-23), Apr. 24-30, May 8-15) (G. Dorner). 2 ♂, 2 ♀. Types, Coll. Barnes.

Similar generically to *viridalis* B. & McD.; agrees in venation with both Ragonot's and Hampson's definition, but lacks the glandular swelling given by Hampson as characteristic of ♂ sex. The green color is very liable to fade in relaxing the specimens, as has happened in the two ♀ specimens before us.

CRAMBINAE

CRAMBUS CHILOIDELLUS sp. nov. (Pl. VIII, Fig. 10.).

Palpi pale ochreous, smoky outwardly; head, thorax and primaries very pale ochreous, almost creamy, all veins lined with smoky brown, interspaces with a broader streak of same color; a fine white streak through cell from base to discal dot, and another streak of same color through entire submedian fold, both bordered on each side by smoky brown; small dark discal dot and terminal row of spots; fringes pale, cut by two transverse smoky lines; secondaries whitish, veins and costa slightly smoky. Beneath primaries and costa of secondaries deep smoky in ♂, paler in ♀, remainder of secondaries whitish. Expanse 25 mm.

HABITAT. White Mts., Ariz. (Lusk). 1 ♂, 2 ♀. Types, Coll. Barnes.

The species looks like a large edition of *densellus* Zell.; in all three specimens however veins 7-9 of primaries are stalked and 10 is from cell, although very close to stalk of 9, 11 is free and the front is non-tuberculate. This would throw the species into *Crambus*, although very dissimilar to the majority of species of this genus. In one ♀ before us the brown streaks are scarcely visible towards base of wing; the white streaks of cell and submedian fold are not prominent as they so nearly coincide with the ground color of the wings.

PLATYTES DENTILINEATELLA sp. nov. (Pl. IX, Fig. 16-17).

Palpi ochreous, smoky outwardly; head and thorax pale ochreous; primaries ochreous, shading into yellow in terminal area; veins and interspaces lined with brown to subterminal line; a brown discal dot, connected with base of wing by white streak; a broader white streak through submedian fold from base to subterminal line, bordered by brown; an indistinct, brown median line, starting from before middle of costa and very strongly outcurved around cell, the outermost portion almost attaining subterminal line, curving backward almost parallel to median vein as far as vein 2 near origin, then with slight outward bend and again oblique to middle of inner margin; submarginal brown line bordered outwardly with pale ochreous, often indistinct, from costa 1/3 from apex rigidly oblique to vein 7 near outer margin, then parallel to same and strongly and evenly dentate in the interspaces; an oblique diffuse dark shade extending inwards from apex to median line; terminal space even yellow, with terminal dark line and interspaceal dots; fringes purplish, cut by a darker line. Secondaries slightly smoky, silky. Beneath primaries smoky, secondaries whitish. Expanse ♂ 18 mm., ♀ 22 mm.

HABITAT. Palmerlee, Ariz. 2 ♂, 4 ♀. Types, Coll. Barnes.

The flat front and veins 7 and 10 of primaries being from the cell agree with Hampson's definition of *Platytes* Gn.; vein 11 of primaries is free and 4 and 5 of secondaries may be either stalked or free. The dentate subterminal line and strong outcurve of median line are characteristic.

PHYCITINAE

MEGASIS RUBRITHORACELLA sp. nov. (Pl. VII, Figs. 7, 9).

♂. Palpi gray, heavily sprinkled with black; tegulae and patagia reddish, central portion of thorax black; abdomen ochreous with a paler ring at the base; primaries dark smoky gray, slightly reddish at extreme base; maculation rather indistinct; t. a. line scarcely paler than ground color, slightly outwardly oblique, straight, followed by a black shade, a pale curved mark extending along median vein and discocellulars; t. p. line pale, inwardly oblique from costa to vein 5, then angled slightly outwardly and minutely dentate on veins 2-5, running subparallel to outer margin, preceded by a blackish shade, strongest at costa; a slight dark shade beyond t. p. line at apex; terminal broken dark line; fringes concolorous. Secondaries evenly pale smoky with paler silky fringes. Beneath pale smoky with traces of t. p. line on costa of primaries, otherwise immaculate.

♀. Smaller and chunkier than ♂; primaries darker and much more clearly and heavily marked; black shades following t. a. and preceding t. p. lines strong and broad; the pale curved mark in cell contains a small discal spot; secondaries deeper smoky. Expanse ♂ 28 mm., ♀ 25 mm.

HABITAT. White Mts., Ariz. (Lusk). 1 ♂, 1 ♀. Types, Coll. Barnes.

Related to *caudellella* Dyar but distinguished among other things but the reddish tegulae and patagia.

BERTELIA Gen. nov. (Type *B. grisella* sp. nov.).

Antennae in ♂ with strong posterior tuft of hair on basal joint, bent, with prominent sinus, but without tuft in same, strongly unipectinate, the pectinations gradually lessening in length towards apex; labial palpi ascending, 3rd joint long, cylindrical; maxillary palpi tufted slightly at extremity, flattened against front; front slightly tufted; primaries 11 veined, 2 and 3 well separate, parallel, 4 and 5 stalked, 8 and 9 stalked, 10 from the cell; secondaries 8 veined, 2 from well before angle, 3 from lower angle, which is considerably produced, 4 and 5 stalked for 1/2 their length, 6 from upper angle, 7 and 8 stalked.

This genus is allied to *Monoptilota* Hulst but lacks the tuft in sinus of ♂ antennae and has a tuft on basal joint not present in Hulst's genus; the pectinations are also longer. Veins 4 and 5 of primaries are stalked in our genus and not separate.

B. GRISELLA sp. nov. (Pl. VIII, Fig. 12).

Palpi, head and thorax gray; abdomen ochreous, paler at base; primaries pale gray, sprinkled heavily with fuscous, maculation indistinct; a faint blackish streak through cell from base to t. a. line, above which costa is paler gray than remainder of wing; t. a. line represented by an outwardly oblique black streak at costa and a pale incurved mark above inner margin, preceded by slight dark shading which in one specimen bends outward to join the costal streak in cell; the t. p. line cannot be traced in either specimen before us, a pale angulate mark on inner margin near angle may represent its termination; there is also some slight dark shading in the subterminal area which cannot be formed into any definite line; fringes concolorous. Secondaries pale hyaline smoky. Beneath pale smoky, basal 1/2 of costa of both wings ochreous. Expanse 24-28 mm.

HABITAT. Redington, Ariz. (Chrisman). 2 ♂. Type, Coll. Barnes.

We figure the co-type which has lost most of the antennae but is the better marked specimen of the two.

YPONOMEUTIDAE

Genus MIEZA Wlk. (*Enaemia* Zell.)

Our present lists, following doubtless Dyar's "Notes on the Yponomeutidae" (Can. Ent. XXXII, 37, 1900) contain under this generic name three species, *subfervens* Wlk. *psammitis* Zell. and *igninix* Wlk. (syn. *crassinervella* Zell.) The species described by Hubner from Georgia (Zutr. Exot. Schmett. III, p. 24, figs. 489-90) as *Eustixis pupula,* which in the older lists was included in this group, has been entirely omitted; we have been unable to find any stated reason for this proceeding and can only surmise that it has either been confused with *Eustixia pupula* Hbn. (Zutr. Exot. Schmett. II, p. 24, figs. 327-8), the well known Pyraustid, or else that it was considered that the figure was unrecognizable and the name dropped. We have examined Hubner's figure and consider that the latter course is certainly justifiable and probably the wisest thing to do under the circumstances. While the figure undoubtedly represents a species of the genus under consideration, the coloration is very crude and faultily applied; the primaries are pale purple-white, the color extending over to the basal portion of secondaries, the cell of primaries being tinged with yellowish; the remainder of secondaries and the under side are red; the two rows of dots on primaries characteristic of the species of this genus are black. The crucial point however is that the veins of *both primaries and secondaries* are outlined in black; all known species have the secondaries uniform red, but one of the main points of difference between *igninix* Wlk. and a second species, common in Southern Texas, is that in the former a large portion of the veins on primaries are distinctly outlined in black, whereas in the 2nd species the venation is not apparent, the whole wing being uniformly white, with the exception of the ordinary dots. It is impossible to determine from Hubner's figure whether these black lines were merely intended to indicate in a general way the venation, or whether they had reference, at any rate in the case of the primaries, to distinct and definite markings; the name *pupula* can therefore, according to the view-point of the investigator, be applied to either of two species with equal justification and should be dropped as 'not sufficiently characterized.' This course of procedure would leave the generic name *Mieza* Wlk. standing, with *Enaemia* Zell. as a synonym, otherwise both would fall before *Eustixis* Hbn. which is sufficiently distinct from *Eustixia* Hbn. to warrant its retention.

With Dyar's synopsis of species we cannot agree; in the first place he separates *subfervens* Wlk. and *psammitis* Zell. on the ground that in the former the basal 2/3 of inner margin of primaries is without brown streaks; if however we turn to the original description we find stated "fore wings white with many elongate brown points" no mention being made of their absence from any portion of the wing. Dyar's diagnosis is evidently based on Stretch's figure (Zyg. and Bomb. N. Am. Pl. 7, fig. 17) and not on Walker's description. Stretch's figure represents a form unknown to us; it may be either an aberrant specimen of *subfervens* or a very worn one in which a portion of the brown scaling has been rubbed off. We consider that *subfervens* Wlk. and *psammitis* Zell. (Pl. IX, Fig. 13) are synonyms, just as stated by Grote (Buff. Bull. II, 152), who had examined Walker's type. Further Dyar unites *igninix* Wlk. and *crassinervella* Zell.; Zeller's figure (Verh. Zool. bot. Ges. Wien. XXII, 563, Pl. III, fig. 27) shows the costal margin of primaries broadly and strongly suffused with dark gray; this is not mentioned in Walker's description nor is it present in any specimens of *igninix* (Pl. IX, Fig. 15) we have seen from Florida, the type locality. We have seen nothing to correspond with *crassinervella*, but in view of the difference in type localities consider that the names should be kept separate; they probably at least represent geograpical races. We suggest therefore the following synonymy.

subfervens Wlk.
> *psammitis* Zell.
igninix Wlk.
crassinervella Zell.

We have recently received from Texas two further apparently new species which we characterize as follows:—

M. BASISTRIGA sp. nov. (Pl. IX, Figs. 11, 12.)

Palpi brick-red, darker towards apex; head and thorax white, the groove of demarcation between the two being finely marked with red, this line extending across the base of the patagia and along the basal portion of costa; mesothorax with one central and two lateral red spots; abdomen brick red with an ochreous ventral stripe; pectus and legs red; primaries rather shiny white with a short red basal streak just below costa and two semi-parallel rows of deep red spots crossing the wings before and after the middle; the antemedial row consists of 3 spots, forming a rather rigid inwardly oblique line, one spot in the cell and the other two below the fold and often more or less conjoined; the postmedial row is parallel to outer margin and is formed by a spot beyond cell and three spots extending obliquely from fold towards inner margin; secondaries pale brick-red with whitish fringes. Beneath both wings brick-red, paler towards inner margin of secondaries. Expanse ♂ 19 mm., ♀ 22 mm.

HABITAT. San Benito, Texas (Mar., May, Sept.), Brownsville, Tex. (G. Dorner). 4 ♂, 5 ♀. Types, Coll. Barnes.

We have an old specimen of this species from Texas labelled by Dr. Dyar *subfervens* Wlk.; it differs however from this species in lacking the brown streaks of primaries and in having a red basal streak below costa; in rubbed specimens this latter is often not apparent.

M. ATROLINEA sp. nov. (Pl. IX, Figs. 9, 10.)

Palpi blackish, red at base; head and thorax white, vertex of head broadly red; base and tips of patagia black; anterior portion of mesothorax and posterior portion of metathorax black; abdomen red with broad ochreous ventral band, legs red outwardly, ochreous inwardly, fore and mid tibiae and tarsi and hind tarsi blackish; primaries shiny white; base of costa and a short subcostal streak at base of wing black; an antemedial oblique broken black line consisting of a streak in the cell, another above the fold and a third between fold and inner margin; a postmedial black line, slightly broken and parallel to outer margin, consisting of an isolated spot beyond cell and a streak extending from origin of vein 6 to just above inner margin; a faint black spot at anal angle; secondaries pale brick-red with whitish fringes. Beneath brick-red. Expanse ♂ 23 mm.

HABITAT. San Benito, Texas (May, Sept.). 8 ♂, 4 ♀. Types, Coll. Barnes.

Easily recognized by the black almost continuous lines in place of the spots of allied species.

PLATE I

Fig. 1. *Argynnis eurynome luski* B. & McD.   White Mts., Ariz.   ♂ Type.
Fig. 2. *Argynnis eurynome luski* B. & McD.   White Mts., Ariz.   ♀ Type.
Fig. 3. *Argynnis eurynome luski* B. & McD.   White Mts., Ariz.   ♂ underside.
Fig. 4. *Argynnis eurynome luski* B. & McD.   White Mts., Ariz.   ♀ underside.
Fig. 5. *Argynnis bischoffi washingtonia* B. & McD.   Paradise Valley, Mt. Ranier.   ♂ Type.
Fig. 6. *Argynnis bischoffi washingtonia* B. & McD.   Paradise Valley, Mt. Ranier.   ♀ Type.
Fig. 7. *Argynnis bischoffi washingtonia* B. & McD.   Paradise Valley, Mt. Ranier.   ♂ underside.
Fig. 8. *Argynnis bischoffi washingtonia* B. & McD.   Paradise Valley, Mt. Ranier.   ♀ underside.

PLATE II

Fig. 1. *Brenthis chariclea ranieri* B. & McD. Paradise Valley, Mt. Ranier. ♂ Cotype.

Fig. 2. *Brenthis chariclea ranieri* B. & McD. Paradise Valley, Mt. Ranier. ♀ Type.

Fig. 3. *Brenthis chariclea ranieri* B. & McD. Paradise Valley, Mt. Ranier. ♂ underside.

Fig. 4. *Brenthis chariclea ranieri* B. & McD. Paradise Valley, Mt. Ranier. ♀ underside.

Fig. 5. *Brenthis chariclea boisduvali* Dup. Nepigon, Ont. ♂ underside.

Fig. 6. *Brenthis chariclea boisduvali* Dup. Calgary, Alta. ♀.

Fig. 7. *Chlorippe clyton subpallida* B. & McD. Babaquivera Mts., Ariz. ♂ Type.

Fig. 8. *Chlorippe clyton subpallida* B. & McD. Babaquivera Mts., Ariz. ♀ Type.

Fig. 9. *Chlorippe clyton subpallida* B. & McD. Babaquivera Mts., Ariz. ♀ underside.

Fig. 10. *Chlorippe clyton texana* Skin. San Antonio, Texas. ♀ underside.

PLATE II

PLATE III

Fig. 1. *Copaeodes rayata* B. & McD. San Benito, Texas. ♂ Type.
Fig. 2. *Copaeodes rayata* B. & McD. San Benito, Texas. ♀ Type.
Fig. 3. *Incita aurantiaca tenuimargo* B. & McD. Redington, Ariz. ♀ Type.
Fig. 4. *Ozodania subrufa* B. & McD. San Benito, Texas. ♂ Type.
Fig. 5. *Ozodania subrufa* B. & McD. San Benito, Texas. ♀ Type.
Fig. 6. *Incita aurantiaca* Hy. Edw. Cal. ♂ typical.
Fig. 7. *Schinia brunnea* B. & McD. Loma Linda, S. Calif. ♂ Type.
Fig. 8. *Schinia brunnea* B. & McD. Loma Linda, S. Calif. ♂ Cotype.
Fig. 9. *Schinia brunnea* B. & McD. Loma Linda, S. Calif. ♀ Cotype.
Fig. 10. *Pygarctia murina albistrigata* B. & McD. San Benito, Texas. ♂ Type.
Fig. 11. *Gortyna serrata* forma *ochrimacula* B. & McD. White Mts., Ariz. ♂ Type.
Fig. 12. *Pygarctia murina albistrigata* B. & McD. San Benito, Texas. ♀ Type.
Fig. 13. *Parastichtis atrosuffusa* B. & McD. White Mts., Ariz. ♀ Type.

PLATE III

152

PLATE IV

Fig. 1. *Grotella spaldingi* B. & McD. Vineyard, Utah. ♂ Type.
Fig. 2. *Grotella spaldingi* B. & McD. Vineyard, Utah. ♀ Type.
Fig. 3. *Pygarctia flavidorsalis* B. & McD. Palmerlee, Ariz. ♂ Type.
Fig. 4. *Crambidia impura* B. & McD. Palmerlee, Ariz. ♂ Type.
Fig. 5. *Crambidia pura* B. & McD. S. Pines, N. C. ♂ Type.
Fig. 6. *Crambidia pura* B. & McD. S. Pines, N. C. ♀ Type.
Fig. 7. *Crambidia dusca* B. & McD. San Diego, Calif. ♂ Type.
Fig. 8. *Tarache dorneri* B. & McD. San Benito, Texas. ♂ Type.
Fig. 9. *Tarache dorneri* B. & McD. San Benito, Texas. ♀ Type.
Fig. 10. *Homocerynea cleoriformis* B. & McD. Palmerlee, Ariz. ♂ Type.
Fig. 11. *Homocerynea cleoriformis* B. & McD. Palmerlee, Ariz. ♀ Type.
Fig. 12. *Givira lucretia* B. & McD. San Benito, Tex. ♂ Type.

PLATE IV

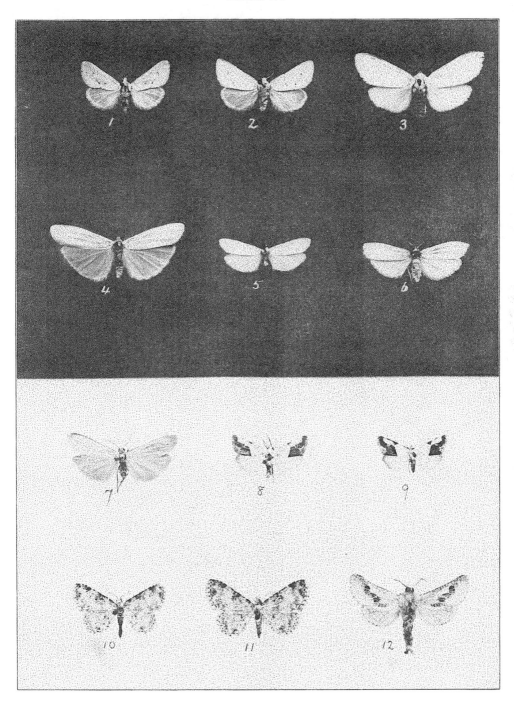

PLATE V

Fig.   1.   *Eriopyga dubiosa* B. & McD.   San Benito, Tex.   ♀ Type.
Fig.   2.   *Monima tenuimacula* B. & McD.   Kerrville, Texas.   ♂ Type.
Fig.   3.   *Eriopyga euxoiformis* B. & McD.   Palmerlee, Ariz.   ♂ Type.
Fig.   4.   *Oncocnemis linda* B. & McD.   Loma Linda, S. Cal.   ♂ Type.
Fig.   5.   *Cucullia minor* B. & McD.   Deming, N. M.   ♂ Type.
Fig.   6.   *Ogdoconta sexta* B. & McD.   Brownsville, Tex.   ♂ Type.
Fig.   7.   *Catabena sagittata* B. & McD.   Loma Linda, S. Cal.   ♂ Type.
Fig.   8.   *Catabena sagittata* B. & McD.   Loma Linda, S. Cal.   ♀ Type.
Fig.   9.   *Proxenus mindara* B. & McD.   Loma Linda, S. Cal.   ♂ Type.
Fig. 10.   *Eustrotia cnossia* Druce.   San Benito, Tex.   ♂.
Fig. 11.   *Eustrotia cnossia* Druce.   San Benito, Tex.   ♀.
Fig. 12.   *Tarache curvilinea* B. & McD.   Brownsville, Tex.   ♂ Type.
Fig. 13.   *Leucocnemis subtilis* B. & McD.   San Benito, Tex.   ♂ Type.
Fig. 14.   *Aleptina inca texana* B. & McD.   Brownsville, Tex.   ♂ Type.
Fig. 15.   *Aleptina inca* Dyar.   Christmas, Ariz.   ♂ typical.

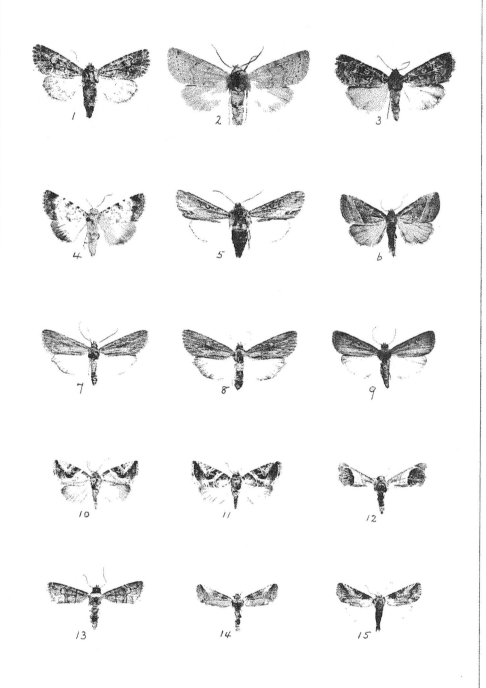

PLATE VI

Fig. 1. *Macaria heliothidata* Gn. Brownsville, Tex. ♂.
Fig. 2. *Macaria heliothidata* Gn. San Benito, Tex. ♀.
Fig. 3. *Macaria heliothidata* Gn. San Benito, Tex. ♂ underside.
Fig. 4. *Macaria ocellinata* Gn. New York. ♂.
Fig. 5. *Macaria ocellinata* Gn. New York. ♀.
Fig. 6. *Macaria ocellinata* Gn. Elkhart, Ill. ♂ underside.
Fig. 7. *Macaria flaviterminata* B. & McD. San Benito, Tex. ♂ Cotype.
Fig. 8. *Macaria flaviterminata* B. & McD. San Benito, Tex. ♀ Type.
Fig. 9. *Macaria flaviterminata* B. & McD. Brownsville, Tex. ♀ underside.
Fig. 10. *Macaria parcata* Grossb. San Benito, Tex. ♂.
Fig. 11. *Macaria parcata* Grossb. San Benito, Tex. ♀.
Fig. 12. *Macaria parcata* Grossb. San Benito, Tex. ♀ underside.
Fig. 13. *Macaria subterminata* B. & McD. San Benito, Tex. ♂ Type.
Fig. 14. *Macaria subterminata* B. & McD. San Benito, Tex. ♀ Type.
Fig. 15. *Macaria subterminata* B. & McD. San Benito, Tex. ♀ underside.
Fig. 16. *Macaria atrimacularia* B. & McD. San Benito, Tex. ♂ Type.
Fig. 17. *Macaria stipularia* B. & McD. San Benito, Tex. ♂ Type.
Fig. 18. *Macaria atrimacularia* B. & McD. San Benito, Tex. ♀ Type.

PLATE VI

PLATE VII

Fig. 1. *Cleora atristrigaria* B. & McD. Brownsville, Tex. ♂ Type.

Fig. 2. *Polia apurpura* B. & McD. White Mts., Ariz. ♂ Type.

Fig. 3. *Cleora atristrigaria* B. & McD. San Benito, Tex. ♀ Type.

Fig. 4. *Xanthorhoe dentilinea* B. & McD. White Mts., Ariz. ♂ Type.

Fig. 5. *Selidosema purpuraria* B. & McD. White Mts., Ariz. ♂ Type.

Fig. 6. *Xanthorhoe dentilinea* B. & McD. White Mts., Ariz. ♀ Type.

Fig. 7. *Megasis rubrithoracella* B. & McD. White Mts., Ariz. ♂ Type.

Fig. 8. *Polia luski* B. & McD. White Mts., Ariz. ♂ Type.

Fig. 9. *Megasis rubrithoracella* B. & McD. White Mts., Ariz. ♀ Type.

Fig. 10. *Diastictis grossbecki* B. & McD. Brownsville, Tex. ♂ Type.

Fig. 11. *Selidosema nigricaria* B. & McD. Palmerlee, Ariz. ♀ Type.

Fig. 12. *Diastictis grossbecki* B. & McD. Brownsville, Tex. ♀ Type.

PLATE VII

PLATE VIII

Fig. 1.] *Therina flavilinearia* B. & McD. Palmerlee, Ariz. ♂ Type.
Fig. 2.] *Euemera immacularia* B. & McD. Loma Linda, S. Cal. ♂ Type.
Fig. 3.] *Therina flavilinearia* B. & McD. Palmerlee, Ariz. ♀ Type.
Fig. 4.] *Synelys vestalialis* B. & McD. Vineyard, Ut. ♀ Type.
Fig. 5.] *Deilinea undularia* B. & McD. White Mts., Ariz. ♂ Type.
Fig. 6.] *Deilinea undularia* B. & McD. White Mts., Ariz. ♀ Type.
Fig. 7.] *Xystrota roseicosta* B. & McD. San Benito, Tex. ♂ Type.
Fig. 8.] *Xystrota roseicosta* B. & McD. San Benito, Tex. ♀ Type.
Fig. 9.] *Scelolophia dorneraria* B. & McD. Brownsville, Tex. ♀ Type.
Fig. 10.] *Crambus chiloidellus* B. & McD. White Mts., Ariz. ♂ Type.
Fig. 11.] *Argyractis confusalis* B. & McD. White Mts., Ariz. ♂ Type.
Fig. 12.] *Bertelia grisella* B. & McD. Redington, Ariz. ♂ Cotype.
Fig. 13.] *Leptomeris benitaria* B. & McD. San Benito, Tex. ♂ Type.
Fig. 14.] *Chalinitis albistrigalis* B. & McD. San Benito, Tex. ♂ Type.
Fig. 15.] *Leptomeris benitaria* B. & McD. San Benito, Tex. ♀ Type.

PLATE VIII

PLATE IX

Fig. 1. *Negalasa rubralis* B. & McD. Redington, Ariz. ♂ Type.
Fig. 2. *Negalasa rubralis* B. & McD. Redington, Ariz. ♀ Type.
Fig. 3. *Negalasa fumalis* B. & McD. San Benito, Tex. ♂ Type.
Fig. 4. *Negalasa fumalis* B. & McD. San Benito, Tex. ♀ Type.
Fig. 5. *Cordylopeza nigripunctalis* B. & McD. Palmerlee, Ariz. ♂ Type.
Fig. 6. *Cordylopeza nigripunctalis* B. & McD. Palmerlee, Ariz. ♀ Type.
Fig. 7. *Penthesilea succulalis* Rag. S. Pines, N. C. ♂.
Fig. 8. *Penthesilea succualis* Rag. Babaquivera Mts., Ariz. ♀.
Fig. 9. *Mieza atrolinea* B. & McD. San Benito, Tex. ♂ Type.
Fig. 10. *Mieza atrolinea* B. & McD. San Benito, Tex. ♀ Type.
Fig. 11. *Mieza basistriga* B. & McD. San Benito, Tex. ♂ Type.
Fig. 12. *Mieza basistriga* B. & McD. San Benito, Tex. ♀ Type.
Fig. 13. *Mieza subfervens* Wlk. (*psammitis* Zell.) Kerrville, Tex. ♂.
Fig. 14. *Illice conjuncta* B. & McD. San Benito, Tex. ♂ Type.
Fig. 15. *Mieza igninix* Wlk. Everglade, Fla. ♂.
Fig. 16. *Platytes dentilineatella* B. & McD. Palmerlee, Ariz. ♂ Type.
Fig. 17. *Platytes dentilineatella* B. & McD. Palmerlee, Ariz. ♀ Type.

PLATE IX

INDEX

Page

albistrigalis B. & McD.......... 136
v. albistrigata B. & McD........ 103
v. alticola B. & McD........... 98
antonia Edw................... 98
aphirape Ochs. 97
apurpura B. & McD........... 106
v. arctica Zett................. 96
arene Edw. 100
arge Stkr. 93
artonis Edw. 94
v. atriclava B. & McD.......... 111
atrimacularia B. & McD....... 127
atristrigaria B. & McD......... 130
atrolinea B. & McD............. 144
atrosuffusa B. & McD.......... 113
aurantiaca Hy. Edw............ 105
basistriga B. & McD............ 143
benitaria B. & McD............ 122
Bertelia B. & McD............. 140
bicolor B. & McD.............. 107
bischoffi Edw. 94
v. boisduvali Dup.............. 96
brunnea B. & McD............. 104
casta Sanb. 101
cephalica G. & R.............. 101
chariclea Schneid 96
chiloidellus B. & McD......... 138
cleoriformis B. & McD........ 119
clio Edw. 94
clyton Bdv. 99
cnossia Druce 117
cocles Lint. 99
columella Druce 121
columelloides B. & McD........ 120
confusalis B. & McD........... 133
conjuncta B. & McD........... 102
Cordylopeza Zell. 134
crassinervella Zell. 142
curvilinea B. & McD. 118
dentilinea B. & McD. 121
dentilineatella B. & McD. 138

dodi Sm. 107
dorneraria B. & McD. 123
dorneri B. & McD. 117
dubiosa B. & McD. 108
dusca B. & McD. 101
elegans Stretch 103
Enaemia Zell. 142
erinna Edw. 93
erratica B. & McD. 113
eurynome Edw. 93
euxoiformis B. & McD. 108
flavidorsalis B. & McD. 103
flavilinearia B. & McD. 131
flaviterminata B. & McD. 125
fumalis B. & McD. 136
Galasa Wlk. 134
gnophosarium Gn. 129
grandimacula B. & McD. 114
grisella B. & McD. 140
grossbecki B. & McD. 128
heliothidata Gn. 124
hepaticaria Gn. 123
hirsutaria B. & McD. 131
Homocerynea B. & McD. 118
igninix Wlk. 142
immaculata B. & McD. 129
impura B. & McD. 101
inca Dyar 115
laceyi B. & McD. 111
Icilia Edw. 98, 99
lignicolora Gn. 112
linda B. & McD. 110
lucretia B. & McD. 132
v. luski B. & McD. 95
macaria Edw. 93
Mieza Wlk. 142
mindara B. & McD. 116
minor B. & McD. 110
v. montis Edw. 98
murina Stretch 103
minuscula B. & McD. 115

	Page			Page
myrtis Edw.	100	rubrothoracella B. & McD.		140
Negalasa B. & McD.	135	sacculalis Rag.		136
nigricaria B. & McD.	129	sagittata B. & McD.		116
nigrinodis Zell.	134	schwarziorum Dyar		102
nigripunctalis B. & McD.	134	sectilis B. & McD.		114
noctiluca Druce	130	serrata Sm.		109
ocellinata Gn.	125	sexta B. & McD.		117
for. ochrimacula B. & McD.	117	*simulata* Hulst.		125
opis Edw.	94	spaldingi B. & McD.		105
ossianus Bdv.	97	stipularia B. & McD.		127
palilis Harv.	108	subfervens Wlk.		142
pallida Pack.	101	subjecta Wlk.		103
pampinaria Gn.	131	subpallida B. & McD.		99
parcata Grossb.	125	subrufa B. & McD.		102
psammitis Zell.	142	subterminata B. & McD.		126
punctolineata Pack.	125	subtilis B. & McD.		114
pupula Hbn.	142	suffusa B. & McD.		101
pura B. & McD.	101	tenuimacula B. & McD.		109
purpuraria B. & McD.	129	*v.* tenuimargo B. & McD.		105
purpurissata B. & McD.	112	*v.* texana B. & McD.		115
v. rainieri B. & McD.	96	*v.* texana Skin.		99
rayata B. & McD.	100	triclaris Hbn.		97
roseicosta B. & McD.	123	undularia B. & McD.		124
rubidana Wlk.	134	vestalialis B. & McD.		122
rubralis B. & McD.	136	*v.* washingtonia B. & McD.		95

TO THE

NATURAL HISTORY

OF THE

LEPIDOPTERA

OF

NORTH AMERICA

VOL. II
No. 4

SOME APPARENTLY NEW LEPIDOPTERA FROM SOUTHERN FLORIDA

BY

WILLIAM BARNES, S. B., M. D.

AND

J. H. McDUNNOUGH, Ph. D.

DECATUR, ILL.
THE REVIEW PRESS
JULY 15, 1913

Published
Under the Patronage
of
Miss Jessie D. Gillett
Elkhart, Ill.

INTRODUCTION

The following species of Lepidoptera were taken on a collecting tour in the spring of 1912 in Southwestern Florida. Mr. J. A. Grossbeck, who also accompanied the expedition, has a complete annotated list of the species captured in course of preparation; this will appear in one of the Bulletins of the American Museum of Natural History.

NOLIDAE

Gelama obliquata sp. nov. (Pl. I, Fig. 2).

♀. Dull gray, primaries crossed by two inwardly oblique parallel smoky lines, the inner rigid, the outer punctiform on the veins and very slightly exserted opposite cell; a pale waved s. t. line, incurved opposite cell and in submedian fold; terminal row of black dots. Secondaries smoky, whitish towards base. Beneath, smoky, secondaries pale at base. Expanse 16 mm.

Habitat. Everglade, Fla. (Apr. 16-23). 1 ♀. Type, Coll. Barnes.

The course of the t. p. line readily distinguishes this species from other N. Am. forms.

NOCTUIDAE

Acidaliodes eoides sp. nov. (Pl. III, Fig. 1).

♀. Head, thorax and primaries light fawn brown, latter with the maculation rather indistinct; t. a. line scarcely traceable, indicated by slight smoky scales, apparently bent outward from costa to below cell and then strongly inwardly oblique to inner margin near base; orbicular indicated by a minute dark dot with slight dark scaling above it on costa; reniform a small black blotch overlaid with paler scales; t. p. line distinct in lower half and strongly oblique inwardly to middle of inner margin, upper half obscure, apparently nearly perpendicular to costa, this line followed by a slightly paler shade than the ground color; s. t. line pale, wavy, parallel to t. p. line, defined outwardly, especially opposite cell, by smoky shading; distinct terminal row of dots; fringes concolorous, sprinkled outwardly with dusky. Secondaries similar to primaries in color and markings, but these latter so indistinct as scarcely to be noticed, leaving merely a slightly mottled appearance; terminal black dots distinct. Beneath pale smoky, silky, primaries rather darker than secondaries. Expanse 12 mm.

Habitat. Everglade, Fla. 1 ♀. Type, Coll. Barnes.

Agrees exactly with Hampson's definition of *Acidaliodes* (Lep. Phal. Brit. Mus. X, 17) except that no tufts are visible on abdomen; the species has a marked resemblance to a small Eoid, but must be precluded from association with the *Geometridae* as vein 5 of primaries is distinctly nearer 4 than 6.

GENUS ARESIA gen. nov. (Type *Aresia parva* sp. nov.).

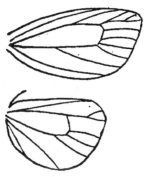

Fig. 1. Venation of Genus *Aresia.*

Antennae in ♂ bipectinate, in ♀ simple; palpi upturned, slender, not attaining vertex of head, 3rd joint slightly bent outwards; proboscis developed; eyes naked, front smooth; thoracic vesture flat, scaly; tibiae unarmed; primaries broad, subquadrate, 11 veined, R_1 from just beyond middle of cell; R_2 wanting, R_3, R_4 and R_5 stalked, areole wanting, M_1 from below angle of cell, M_2 bent downwards towards M_3 at base, M_3 from above lower angle of cell, Cu_1 from angle, Cu_2 from middle of cell; secondaries 7 veined, S.C. joined with median at base only, R_1 and M_1 stalked, M_2 from below middle of cell, well-developed, M_3 absent, Cu_1 from angle of cell, Cu_2 well before angle.

The genus belongs apparently in Hampson's subfamily *Erastrianae;* it has certain affinities in general appearance with the *Sarrothripinae* but lacks the bar-like retinaculum and raised scales on primaries.

A. PARVA sp. nov. (Pl. I, Fig. 1).

Head and thorax whitish; primaries whitish, suffused with pale green, and shaded with light purple; basal 1/3 of wing very pale whitish green, costa with dark dot at base and angulate black line about 1/5 from base of wing, reaching only to cell; t. a. line single, black, rather rigid, slightly outwardly oblique with slight angles at costa and above inner margin; median space purplish, deepest along t. a. line; t. p. line single, black, irregular, angled inwardly below costa and outwardly below cell, in general parallel to t. a line; costal portion of wing beyond t. p. line pale greenish with several purple costal dots, sharply defined below by an oblique line proceeding from apex of wing to opposite cell; below this line the remainder of wing is purplish with the exception of a greenish tinge bordering t. p. line; a faint whitish waved s. t. line is evident crossing this purple color preceded by a dark dot opposite cell; terminal dotted dark line; fringes dusky. Secondaries pale smoky, darker outwardly with a diffuse discal dot. Beneath largely smoky, white at base of secondaries. Expanse 12.5 mm.

HABITAT. Everglade, Fla. (Apr. 1-7). 1 ♂. Type, Coll. Barnes.

We possess several similar specimens from Brownsville, Texas, slightly more contrasting in coloration and with the lines more prominent, especially the subbasal one; we consider them, however, to belong to the same species.

PROROBLEMMA TESTA sp. nov. (Pl. IV, Fig. 5).

Primaries brownish, shading into purplish outwardly, sprinkled slightly with minute black atoms; maculation indistinct; orbicular and reniform represented by slight black dots in the cell; t. p. line faint, pale purple, accentuated by black opposite cell, bent outwards below costa, strongly incurved below cell and waved to origin of vein 2, then oblique to middle of inner margin, ending in a dark dot; s. t. line pale purple, indistinct, waved, close to t. a. line in central portion and parallel to same, bent outwards to anal angle below fold; terminal row of minute black dots; fringes concolorous. Secondaries smoky, whitish towards base with terminal row of black dots. Beneath, smoky, with terminal row of dots to both wings, secondaries paler. Expanse 17 mm.

HABITAT. Everglade, Fla. (Apr. 1-7). 1 ♀. Type, Coll. Barnes.

The wing shape reminds one of a minute specimen of the genus *Phiprosopus* Grt., the venation and other characters agree however with Hampson's genus *Proroblemma* (Cat. Lep. Brit. Mus. X, 34) in which genus we place it for the present.

TYRISSA MULTILINEA sp. nov. (Pl. II, Fig. 5).

Palpi smoky brown, paler beneath, frontal tuft deep brown below, grayish above, head and thorax purplish gray, suffused with brown, especially the tegulae, which are crossed by a black-brown line near apex; abdomen pale ochreous; primaries angled on vein 4, pale gray-brown, heavily suffused with deep purplish and whitish, giving a general mottled appearance; basal line single, black, forming prominent outward angles below costa and on cubital vein; beyond it a reddish-brown costal patch; subbasal space crossed by several indistinct waved diffuse lines; t. a. line reddish-brown, irregularly dentate from costa to cubitus, below which it curves strongly inward to inner margin 1/3 from base, slightly thickened just below cubitus, preceded, especially at costa, by pale shade-line; a diffuse smoky median shade, rounded outwardly across cell, then approaching close to t. a. line and parallel to same from cubitus to inner margin; beyond this some rather prominent whitish shading, especially beyond cell; space between median shade and t. p. line crossed by three equidistant and parallel waved shade-lines; t. p. line distinct, red-brown, bent outward gently below costa, then parallel to outer margin, very faintly spotted with white on the veins; s. t. line indistinct, red-brown, in general parallel to t. p. line; costal portion of subterminal and terminal spaces gray with small brown apical ocellus; remainder of space heavily shaded with deep purplish; dark terminal line; fringes smoky with pale basal line; secondaries with the lines of primaries continued and with a rather prominent oval discal ringlet. Beneath pale ochreous, primaries with dark costal dot at base of wing, a black dot in cell followed by a larger black ring representing orbicular and reniform; between these a dark median line; t. p. line represented by a series of minute black scallops, slightly broken by the paler veins; secondaries with median and t. p. lines as on primaries and with a very prominent black ringlet resting on median line; terminal space rather heavily shaded with black. Expanse 22-25 mm.

HABITAT. Everglade, Fla. (3 ♀); San Benito, Texas, (1 ♂). Types, Coll. Barnes. Cotype, British Museum.

The species has been referred for us by Sir. Geo. Hampson to the genus *Tyrissa* Wlk.; he writes us that it is allied to *recurva* Wlk.

ANOMIS SERRATA sp. nov. (Pl. II, Fig. 1).

♂. Antennae very strongly serrate and fasciculate, almost bipectinate; primaries with the coloration brighter and more contrasting than in *erosa* Hbn., with similar maculation; basal half of wing to a line formed by inner edge of reniform and lower portion of t. p. line bright yellow, heavily sprinkled with orange-brown; beyond this line darker brown, suffused heavily with purple subterminally, this color extending backward along inner margin to base of wing; ordinary lines reddish brown, basal 1/2 line angled outwardly on cubitus; t. a. line outwardly oblique, from costa 1/4 from base to below cubitus, then angled and slightly concave to middle of inner margin; orbicular a white dot narrowly ringed with brown; reniform partially hidden by dark shading of wing, upright, resting on bend of t. p. line, outlined in blackish with dark central shade; t. p. line bent outward slightly at costa, straight to vein 5, angled sharply outwardly on vein 4, bent backwards along vein 3 to its origin, slightly convex to inner margin where it joins t. a. line; s. t. line defined by difference between purple subterminal shade and brown terminal one, irregular, dentate on veins 3 and 4; fringes dark inwardly, whitish and slightly checkered outwardly, not so prominently scalloped as in *erosa*. Secondaries smoky, paler towards base, without the ochreous tinge found in *erosa*. Beneath primaries smoky, slightly pinkish, paler along outer border; t. p. line defined at costa, followed by dark triangular costal patch; faint pale lunate discal spot; secondaries pale ochreous, costal half sprinkled with red-brown, with fine, brown, distinctly undulate postmedian line extending half across wing and pale discal mark.

♀. paler than the ♂, more as the ♂ *erosa* in color. Expanse 28 mm.

HABITAT. Chocoloskee, Fla.; Ft. Meade, Fla. (Apr.).. 2 ♂, 1 ♀. Types, Coll. Barnes.

We should have hesitated to separate this from *erosa* Hbn. if it had not been for the strongly serrated antennae of the ♂; it is smaller in size, brighter in color and the line on underside of secondaries is scalloped, not straight, as is usually found in *erosa*.

We have a third ♂ specimen very worn labeled Glenwood Spgs. Colo., which may possibly be an error for Glenwood, Fla.

PSYCHIDAE

Manatha nigrita sp. nov. (Pl. IV, Fig. 3).

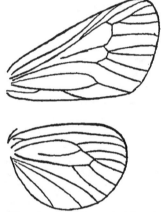

Size and general color of *Eurycyttarus confederata* G. & R. but slightly duller black and eyes much larger. Primaries 12 veined, 1 b and 1 c anastomosing, former with single branch to inner margin, 4 and 5 either connate or shortly stalked, 8 and 9 stalked, 7 either from a point with 8 and 9 or shortly stalked, 10 and 11 free from cell, 11 occasionally anastomosing with 12; secondaries with 8 veins, 4 and 5 either connate or stalked, apparently no crossbar between 7 and 8. Expanse 16-18 mm.

Habitat. Everglade, Fla. (Apr. 8-15). Ft. Myers, Fla. (Apr. 24-30). 7 ♂. Type, Coll. Barnes.

Fig. 2. Venation of *Manatha nigrita*.

The species may be at once distinguished from *confederata* G. & R. by the venation, all veins being present. As is usual in this family, considerable variation is evident in the point of origin of the various branches; our Everglade specimens all show vein 7 of primaries from a point with 8 and 9; in one of the Ft. Myers specimens it is shortly stalked and a long series before us from Brownsville, Texas, of what is apparently the same species agrees with this latter. Hampson's definition and figure of the genus *Manatha* Moore (Moths Brit. Ind. I. 297/8) would cover our species with the exception that *nigrita* lacks the cross-bar between veins 7 and 8 of secondaries in all specimens examined. Our species cannot be *edwardsii* Heyl. as that author distinctly states in the description that the secondaries have only 7 veins.

The sacks were found quite abundantly in both Everglade and Ft. Myers on the dry stalks of last year's grass and from these most of our specimens were bred. The sack is very similar to that of *confederata* but is much finer in quality, being composed of small pieces of grass firmly woven together, the whole thatched with short longitudinal pieces of the grass-stems. The larvae evidently feed on coarse grasses.

GENUS PROCHALIA gen. nov. (Type *Prochalia pygmaea* sp. nov.).

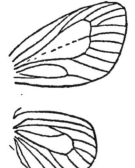

Antennae of ♂ bipectinate; fore tibiae with long spine; primaries 12 veined, 1b anastomosing with 1c but without branch to inner margin, veins 2-6 well separated, 7 and 8 stalked, 9 and 10 from around upper angle of cell, 11 from near middle of cell; secondaries 8 veined, 4 and 5 from a point, 7 and 8 not joined by a crossbar.

The genus falls in Hampson's Subfamily *Chaliinae* in lacking the branch of vein 1b to inner margin; characteristic is the stalking of veins 7 and 8 of primaries, 9 being free.

Fig. 3. Venation of Genus *Prochalia*.

PROCHALIA PYGMAEA sp. nov. (Pl. III, Fig. 5).

♂. Head, thorax, abdomen and wings uniform black-brown, much as in *M. nigrita* B. & McD. Abdomen extending slightly beyond margin of secondaries. Expanse 12.5 mm.

HABITAT. Everglade, Fla. (May 1-7). 1 ♂. Type, Coll. Barnes.

The sacks are narrow, cylindrical tubes of silk 12 mm. long, mixed with excrement and more or less covered with small pieces of lichen. They were not rare on the trunks of orange trees, feeding doubtless on the lichens which covered the bark; only one specimen was, however, successfully reared, this emerging after leaving Everglade. It is probably this species referred to by Hylaerts, (Bull. Soc. Ent. Belg., 1884, p. 209), as an unknown sack from Florida.

COCHLIDIIDAE

PROLIMACODES SCAPHA ARGENTIMACULA subsp. nov. (Pl. II, Fig. 2).

Differs from typical *scapha* in having the silvery line from apex of wing closer to outer margin and the basal angle of same filled with silvery white, nearly as in *trigona* H. Edw.

HABITAT. Ft. Myers, Fla. (Apr. 16-23, May 1-7). 6 ♂. Type, Coll. Barnes.

This constant form seems worthy of a varietal name; it probably represents a geographical race.

PYRAUSTINAE

GLAPHYRIA BASIFLAVALIS sp. nov. (Pl. IV, Fig. 12).

Palpi and thorax white, the former ringed with yellow-brown, the latter sprinkled with yellow; primaries white, heavily shaded with smoky brown, so that very little of the ground color is visible; a pale yellow dash from base of wing to antemedial line through fold; anterior lines evenly rounded, pale, edged with smoky and with light brown shading before and after on costa; a slight dark discal spot; postmedial line bent outwards around cell, followed on costa by light brown shading which tends to deepen and form a quadrate spot opposite cell; two white triangular patches on terminal border, one below apex, the other above the fold; fringes brownish, slightly cut by white. Secondaries smoky, paler towards base and inner margin with faint postmedial line, angled inwards at vein 2; basal half of fringe brown, outer portion whitish, slightly sprinkled with brown scales; long white fringes at anal angle. Beneath primaries smoky, base of wing above inner margin white, apex brown; slight white dashes along outer margin; fringes checkered brown and white; secondaries white with brownish shading along costa and terminal border, a brown discal dot, postmedian curved line and terminal dots, fringes as above. Expanse 11-14 mm.

HABITAT. Everglade, Fla.; Chocoloskee, Fla.; San Benito, Texas. 6 spec. Type, Coll. Barnes.

None of our specimens are very perfect but the yellow basal dash seems sufficiently characteristic to distinguish the species from any described form. There is a tendency for veins 10 and 11 of primaries to entirely coalesce.

SYLEPTA MASCULINALIS sp. nov. (Pl. II, Fig. 10).

♂. Antennae with a strong flattened tuft of scales on inner side of basal joint; fore tibiae very short, broadened apically, with tuft of hair; base of primaries with slight costal tuft of scales on upper side; primaries deep shiny smoky-brown, crossed by two pale yellow lines; the anterior line is evenly rounded outwardly, edged outwardly with darker shade-line attaining inner margin 1/3 from base; orbicular and reniform two elongate black spots separated by a quadrate patch of pale whitish ochre which does not extend either above or below the spots; t. p. line expanded to small patch on costa, curved evenly outward to vein 3, then strongly elbowed inwardly to near base of vein 2, then again straight and slightly waved to inner margin 2/3 from base, edged inwardly by a dark shade; dark terminal shade-line and pale basal line to slightly checkered fringes. Secondaries slightly deeper in color than primaries, t. p. line continued across wing, slightly waved, with strong inward bend along vein 2; dark dash in cell; terminal line and fringes as on primaries. Beneath much paler than above with maculation indistinctly visible.

♀. Similar to ♂ without basal tufts to antennae and costa of primaries; veins of primaries slightly outlined beyond cell in ochreous. Expanse 18 mm.

HABITAT. Chocoloskee, Fla. (♂), Marco, Fla. (Apr. 16-23) (♀). 1 ♂, 1 ♀. Types, Coll. Barnes.

The species apparently belongs close to Hampson's Section IV (*Pramadea*) of this genus. Vein 7 of primaries is not as strongly curved nor so approached to vein 8 as is usual, the species being intermediate between *Sylepta* and *Lygropia* in this respect.

NACOLEIA HAMPSONI sp. nov. (Pl. IV, Fig. 2).

Palpi white beneath, fuscous above suffused with ochreous; fore legs banded with black, front fuscous, thorax largely yellow-brown, shaded with fuscous, abdomen yellow, fuscous towards anal extremity; primaries brownish yellow with broad terminal black-brown border narrowing towards tornus, a median band of similar color, expanding to form a large square patch on costal half of wing with central yellow spot; costa from base to median band black-brown, base of wing very slightly blackish; fringes dusky with yellow patch at anal angle; secondaries similar in ·color to primaries with black-brown border narrowing towards anal angle and a slightly irregular antemedial band of varied width; fringes yellow, smoky towards apex. Beneath, paler yellow than above with markings repeated except that the medial band of primaries is confined to the square costal patch and that of secondaries is lacking, a small discal spot being present. Expanse 19 mm.

HABITAT. Marco, Fla.; Chocoloskee, Fla. 3 ♀. Type, Coll. Barnes. Cotype, British Museum.

Sir Geo. Hampson, to whom we sent a specimen writes us that the species is unknown to him; we take pleasure therefore in naming the species after him.

LOXOSTEGE ALBICEREALIS FLORIDALIS subsp. nov. (Pl. II, Fig. 3).

Ground color of primaries with less of a bluish tinge than in typical *albicerealis;* costal yellow portion not so bright and more heavily shaded with chocolate brown outwardly; whitish patch at base of veins 2 and 3 distinct; dark basal portion of fringe cut with pale ochreous very distinct. Expanse 27 mm.

HABITAT. Everglade, Fla. (Apr. 26-30). 4 ♂, 4 ♀. Type, Coll. Barnes.

The larvae were quite common on a shrub with small narrow, very fleshy leaves, called by· the natives "Florida Cranberry"; they bear considerable resemblance to those of *Tholeria reversalis.* As all our specimens are very constant in coloration we consider a varietal name advisable.

MICROCAUSTA FLAVIPUNCTALIS sp. nov. (Pl. I, Fig. 4).

Primaries brown crossed by two parallel darker lines, curved outward below costa and then inwardly oblique to inner margin; an orange dot at end of cell and apical margin of wing shaded with orange. Secondaries whitish with a large orange patch in cell; outer margin slightly smoky. Beneath, whitish, slightly smoky, with traces of a smoky, submarginal line on secondaries. Expanse 9 mm.

HABITAT. Ft. Myers, Fla. (Apr. 1-7). 1 ♂. Type, Coll. Barnes.

The markings are similar to those of *ignifimbrialis* · Hamp. but our new species is at once recognized by the orange patch on secondaries. If it had not been for this very distinctive feature we had scarcely dared describe the species as our specimen is decidedly worn. Structurally it belongs without a doubt in the genus *Microcausta* Hamp.

STENOPTYCHA SOLANALIS sp. nov. (Pl. IV, Fig. 6).

Palpi covered with purplish-ochreous and blackish scales; head and thorax largely pale purplish admixed with ochreous; abdomen blackish, admixed with ochreous and with basal segments pale ochreous with a dark band at extreme base of abdomen; primaries deep purplish, slightly iridescent, especially along inner margin; outer portion of costa to near apex rather broadly ochreous with 3 or 4 dark spots; maculation indistinct; in the fold about 1/2 way from base to anal angle is an elongated claviform mark edged outwardly with pale ochreous; at end of cell is a lunate dark reniform mark, more or less open at top and bottom, filled with pale ochreous and edged slightly outwardly with same color; a dark postmedian line arising from costa between spots 2 and 3 and very oblique outwardly crossing the ochreous space, then sharply angled, slightly sinuate and subparallel to outer margin to a point below reniform where it stops, not attaining inner margin; a pale line very close to outer margin followed at anal angle by a dark patch; fringes dark, white patch above anal angle. Secondaries smoky, broadly paler along inner margin; dark scale patch beyond cell; inner margin scaled with black; fringe whitish, blackish apically. Beneath, smoky, costa ochreous; markings of upper side very slightly visible; secondaries much as above. Expanse 19 mm.

HABITAT. Everglade, Fla. (Apr. 16-23). 2 ♀. Type, Coll. Barnes.

The species was bred from larvae collected on *Solanum* sp. along with *L. integra* Zell. The genus has not yet been recorded from N. America; our specimens agree in venation with Hampson's figure.

NYMPHULINAE

PILETOCERA SIMPLICIALIS sp. nov. (Pl. IV, Fig. 1).

Antennae of ♂ normal and ciliate, no fovea in cell of primaries; head, thorax and wings deep blackish brown; primaries with median area slightly paler; t. a. line even, pale, slightly rounded, followed by a broad diffuse dark shade which extends across the small round orbicular, more or less concealing it; reniform rather large, subquadrate, black with pale central dash; t. p. line, pale, very irregular, angled outwardly on vein 7 and then very strongly inwardly, almost attaining reniform, rounded outwardly between veins 2 and 5, bent back to below reniform and then straight to inner margin, preceded by a broad black shade, especially prominent near costa and inner margin. Secondaries blackish with faintest trace of a pale subterminal line; fringes of both wings dusky. Beneath, primaries dusky with faint traces of markings of upper side apparent; secondaries with two black spots in cell and a faint s. t. line very prominently excurved between veins 3-5. Expanse ♂, 15 mm., ♀, 18 mm.

HABITAT. Chocoloskee, Fla., ♂; Everglade, Fla., ♀ (Apr. 8-15). 1 ♂, 2 ♀. Types, Coll. Barnes.

Very similar to *bufalis* Gn. but the ♂ lacks the fovea and distorted membrane in the cell of primaries and the t. p. line in both sexes shows much deeper indentations. It is a slightly narrower winged species.

PYRALINAE

HERCULIA SORDIDALIS sp. nov. (Pl. II, Fig. 12).

Pale straw-color heavily sprinkled with smoky, giving a slightly greenish appearance; costa ochreous to just beyond t. p. line, slightly pinkish at base; wings crossed by two pale lines, heavily bordered with smoky which is more prominent than the pale color; primaries with lines wide apart, anterior line evenly rounded with outward smoky border; posterior line almost rigidly perpendicular to inner margin with inward dark border; faint discal dot. Secondaries with lines approached, almost contiguous at inner margin with slightly darker median space; fringes of both wings dusky with fine ochreous basal line. Beneath paler than above with very indistinct maculation. Expanse 25 mm.

HABITAT. Ft. Myers, Fla. (Apr. 1-7). 1 ♀. Type, Coll. Barnes.

Related to *thymetusalis* Wlk. but distinguished by the lack of costal blotches.

MACROTHECINAE

MACROTHECA UNICOLORALIS sp. nov. (Pl. III, Fig. 6).

Head, thorax and primaries dark gray, latter almost unicolorous immaculate, basal 1/3 of wing slightly deeper in color, due to a sprinkling of black scales; minute discal dot and slight dark terminal line; fringes concolorous. Secondaries whitish slightly tinged with smoky. Beneath paler than above. Expanse 10 mm.

HABITAT. Everglade, Fla. 1 ♂. Type, Coll. Barnes.

SCHOENOBIINAE

PATISSA SORDIDALIS sp. nov. (Pl. IV, Fig. 4).

♂. Head, thorax and abdomen pale brownish, latter with a dorsal black patch on 3rd segment and lateral dark shading posteriorly; primaries pale brownish crossed by two smoky blackish lines, the antemedial rounded outwardly, the postmedian in outer fourth of wing, bent slightly at costa, then rigid and inwardly oblique; faint dark discal dot and terminal line. Secondaries paler than primaries with dark lines continued, the inner one being somewhat interrupted. Beneath paler than above with traces of markings of upper side. Expanse 17 mm.

HABITAT. Everglade, Fla. (Apr. 8-15). 1 ♂. Type, Coll. Barnes.

The species is very distinct from any species of this genus known to us. Our specimen is somewhat worn and it is probable that fresh specimens will show the banding more clearly.

PATISSA FLAVIFASCIALIS sp. nov. (Pl. III, Fig. 2).

♂. Head and palpi white, latter marked with light brown outwardly; thorax and abdomen white sprinkled with golden-yellow; primaries white, sprinkled with brown atoms and crossed by three broad yellow bands; the first, at base of wing, is broadest at costa, narrow or broken in central portion, outer margin irregular; the second, slightly before middle of wing, is rather evenly rounded outwardly, slightly narrowed in cell, bordered on both sides by brown sprinkling; between bands 1 and 2 is a slight yellowish patch in cell overlaid with brown atoms; band 3 is subterminal, broadening towards inner margin, upright, bordered with brown dots, connected from the middle of its inner margin with a yellow discocellular patch; beyond this band is a rather prominent line of ground color, followed by four broad yellow transverse streaks extending to outer margin between apex of wing and vein 4; brown terminal line; fringes white slightly cut by brown. Secondaries similar in color to primaries with yellow patch in cell joined to a broad yellow band representing band 3 of primaries, three terminal yellow dashes as on primaries;

inner margin spotted with yellow above anal angle with prominent white scale-tuft tipped with yellow; border and fringes as on primaries. Beneath suffused with smoky with postmedial smoky band crossing both wings. Expanse 11 mm.

HABITAT. Ft. Myers, Fla. (May 1-7). 1 ♂. Type, Coll. Barnes.

The species belongs in the *flavicostella* group. The single specimen was taken in the day-time in a grassy pasture beneath pine trees.

CRAMBINAE

THAUMATOPSIS FLORIDELLA sp. nov. (Pl. II, Fig. 9).

Antennae in ♂ unipectinate; palpi smoky; head and thorax pale ochreous; primaries ochreous brown, basal portion of costa darker; from base of wing through cell extends a narrow white line, broadening gradually to postmedian line, beyond this becoming a diffused, broadish, rather strigate shade extending to outer margin below apex; this line is bordered costally from middle of cell to apex of wing by a fine black broken line; above this black line a deeper shade of brown than general ground color extends to apex of wing from middle of cell, defined faintly with whitish on its upper edge; a white streak through submedian fold from base to submedian line; inner margin broadly white sprinkled with fuscous; branches of cubitus and median in submedian area faintly white; from beyond the cell a dotted smoky postmedian line extends obliquely inwards to middle of inner margin, particularly well marked on white areas of wing; this is followed by a similar semi-parallel subterminal line; terminal row of black dots; fringe dusky, slightly metallic at base. Secondaries white, in ♂ with costal and apical portions broadly smoky, in ♀ with mere traces of smoky terminal band. Beneath, primaries smoky, veins slightly streaked with white terminally, secondaries as above. Expanse ♂ 23 mm., ♀ 30 mm.

HABITAT. Everglade, Fla. (Apr. 1-15); Marco, Fla. (Apr. 16-23). 5 ♂, 4 ♀. Types, Coll. Barnes.

The specimen from Key West doubtfully referred to *fernaldella* by Kearfott (Can. Ent., 37, 122) is probably this species. Our series shows that it is sufficiently and constantly distinct from the above species to warrant a name.

PLATYTES PUNCTILINEELLA sp. nov. (Pl. II, Fig. 11).

Primaries light fawn-brown, slightly sprinkled with black; veins bordered on both sides with white lines with the exception of the portion between submedian fold and inner margin which is unicolorous brown; a fine white streak through the cell to outer margin containing a black dot at end of cell and a fainter one beyond in subterminal space; a distinct subterminal row of black dots close to outer margin and parallel to same; a similar row of terminal dots; fringes concolorous. Secondaries in ♂ smoky, paler towards base, in ♀ whitish with slight smoky suffusion outwardly. Beneath, primaries smoky in ♂, pale ochreous in ♀; secondaries as above. Expanse ♂ 24 mm., ♀ 29 mm.

HABITAT. Everglade, Fla. (Apr. 8-15). 3 ♂, 4 ♀. Types, Coll. Barnes.

Allied to *multilineatella* Hlst., differing from this and allied species in the distinct row of subterminal black dots of primaries.

GENUS STORTERIA gen. nov. (Type *Storteria unicolor* sp. nov.).

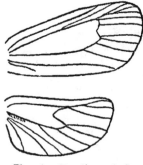

Fig. 4.   Venation of Genus *Storteria.*

Antennae strongly flattened laterally, slightly annulate; labial palpi minute, upturned, not attaining front; maxillary palpi distinct, tufted; proboscis minute, front not prominent, tibiae with well-developed spurs, outer spurs 2/3 the length of inner ones; primaries broad, subquadrate; 12 veined, 4 and 5 close together from lower angle of cell, 6 from slightly below upper angle, 7 from angle, connate with 8, 9 and 10 which are stalked, 11 free from middle of cell; secondaries with cubitus pectinate on upper side, 8 veined, 4 and 5 connate, 6 from upper angle, 7 and 8 stalked.

The genus will fall somewhere near *Leucargyra* Hamp. and *Doratoperas* Hamp. Characteristic is the stalking of veins 8, 9 and 10 of primaries with vein 11 free. The minute palpi readily separate it from our N. American genera.

STORTERIA UNICOLOR sp. nov. (Pl. II, Fig. 4).

♂. Head, thorax, and primaries unicolorous pale seal-brown, silky, with a slightly deeper shade in the cell; secondaries deep smoky, paler at base and along inner margin. Beneath, smoky with costa of primaries dull ochreous. Pro and midlegs deep silky brown inwardly, hind legs pale brown, palpi ochreous, brown outwardly. Expanse 29 mm.

HABITAT. Everglade, Fla. (Apr. 8-15). 1 ♂. Type, Coll. Barnes.

179

ENDOTRICHIINAE

GENUS DAVISIA gen. nov. (Type *Davisia singularis* sp. nov.).

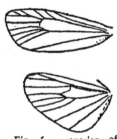

Fig. 5. venation of Genus *Davisia*.

Palpi slender, upturned, not exceeding front; maxillary palpi short, filiform; antennae laterally flattened, slightly annulate and ciliate; proboscis present, but reduced; hind tibiae with two pairs of spurs of equal length; primaries 12 veined, vein 2 from before cell, 3 and 4 from a point, 5 slightly removed from 4 at base, 6 from below upper angle, 7, 8 and 9 stalked, 9 from 8 before 7, 10 and 11 free and parallel; secondaries 8 veined, 2 from before cell, 3 and 4 from a point, 5 from slightly above, 4, 6 from upper angle, 7 and well stalked, median nervure on upper side nonpectinate.

According to Hampson's table (Proc. Zool. Soc. Lond. 1898, p. 591) the genus would fall into the subfamily *Endotrichiinae,* of which so far no American species are recorded. It might possibly be regarded as an aberrant genus of *Epipaschiinae* without tufts on primaries or basal tuft to antennae; in general appearance the type species resembles a small *Noctuelia*. We take pleasure in naming the genus after our friend Mr. W. T. Davis of Staten Is., who formed one of the party in Florida.

DAVISIA SINGULARIS sp. nov. (Pl. III, Fig. 3).

Thorax ochreous, shaded with brown; primaries various shades of brown; black dot at base of costa; basal portion of wing light brown, defined outwardly by a smoky t. a. line 1/4 from base bordered inwardly with pale ochreous and slightly angled outwardly in cell; beyond this line is a brown band of similar color and width to basal space, the outer margin of which is evenly outwardly oblique; the remainder and larger portion of the median space has a pale ochreous ground color, shaded heavily with smoky outwardly, most prominent and broadest above inner margin; a small black discal dot, t. p. line closer to outer margin than usual, blackish, followed by pale shade evenly rounded and slightly dentate from costa to vein 2 where it forms a prominent inward angle and continues perpendicular to inner margin; terminal space light brown very slightly shaded with smoky at apex and above anal angle; terminal broken dark line; fringes dusky at base, paler outwardly. Secondaries smoky with a faint pale subterminal line, most distinct towards anal angle where it is shaded inwardly with smoky; fringes pale. Beneath, primaries deep smoky, secondaries as above. Expanse 11 mm.

HABITAT. Everglade, Fla. (Apr. 8-15). 5 ♂. Type, Coll. Barnes.

EPIPASCHIINAE

Jocara perseella sp. nov. (Pl. II, Fig. 6).

Palpi ochreous, shaded outwardly with greenish and purplish; antenna! tuft ochreous; thorax an admixture of pale green and ochreous scales; abdomen ochreous; primaries a mixture of reddish purple, ochreous and pale green shades; the basal portion of costa is scaled with very broad flat scales of all three colors, beyond this scaling two small dark blotches separated by ochreous and beyond this a large semitriangular ochreous patch, extending to just before t. p. line and separated from same by a reddish purple streak; t. a. line not defined; slight dark point at end of cell; t. p. line pale ochreous, sharpest at costa, gently outcurved between veins 2 and 8, slightly dentate on veins, preceded by a diffuse reddish purple shade; terminal space even greenish with slight dark apical dash and prominent row of black terminal spots; fringes pinkish cut by smoky at base. Secondaries whitish at base and along inner margin, remainder of wing deep smoky with darker broken terminal line; fringes tinged with rosy, with pale ochreous line at base. Beneath pale ochreous, largely suffused with pinkish outwardly and along costa, small discal dot on both wings and traces of pale subterminal line, best marked at costa of primaries. Expanse 22 mm.

Habitat. Everglade, Fla. (Apr. 16-23). 1 ♂. Type, Coll. Barnes.

Several larvae of this very distinct and peculiarly colored species were collected from Alligator Pear (*Persea gratissima*) but only a single perfect specimen was reared; the larvae, of which, unfortunately, no description was made, are found singly or in twos and threes in a slight web on the young tender leaves; they remind one somewhat of the larvae of *Atteva aurea*.

Tetralopha querciella sp. nov. (Pl. II, Figs. 7, 8).

Head and thorax gray; abdomen pale ochreous, ringed with smoky at base of segments; primaries pale gray at base, shading gradually into pale purplish-brown towards t. a. line, this color being most prominent beyond a raised blackish scale patch, extending between cubital vein and inner margin; costa with black point at base; t. a. line outwardly oblique to middle of inner margin, slightly rounded, in ♂ indented by the furrow in the cell, blackish, geminate, filled with ground color; median space pale gray, tinged with brown along t. p. line; in ♂ a prominent brownish patch on costa above the furrow; in ♀ this is wanting, but this latter sex has a black discal point in a slightly raised patch of scales; raised scale-band from inner margin to cell, not prominent, owing to color being the same as that of the surrounding area; veins beyond cell slightly marked in black, more prominent beyond t. p. line and in ♀ sex; t. p. line black, with prominent outward bulge between veins 3-5 and very slight outward angle on vein 1, edged outwardly by a pale gray line which is followed by a diffuse purple-brown shade, broadest at apex; terminal area gray with broken black terminal line, fringes gray. Secondaries even smoky, scarcely paler towards base; fringes cut by a smoky basal line. Beneath, primaries smoky with costal tuft of ♂ paler; faint traces of t. p. line of upper side visible, especially on costa where it is followed by a pale spot; secondaries paler, rather ochreous, with dark spot near base of wing and faint smoky dentate postmedian line, not attaining inner margin; both wings with terminal dark line and pale fringes. Expanse 19-22 mm.

HABITAT. Marco, Fla. (May 15-31). 2 ♂, 4 ♀. Types, Coll. Barnes.

This species was bred from larvae found webbing the terminal twigs of scrub oak; it is close to *floridella* Hlst. from which it can readily be separated by the color of secondaries. It is also related to *melanogrammos* Zell. but differs in several details from the description, especially in the darker secondaries; if Hulst has been correct in his identifications the larvae of Zeller's species are pine feeders, which would point to distinct species. Hulst, however, is noted for his poor determinations and until the types have been carefully studied some element of doubt must exist; we consider it wise, however, to fix this oak-feeder with a name. We follow Hampson in our generic reference.

PHYCITINAE

ACROBASIS TENUELLA sp. nov. (Pl. IV, Fig. 8).

♂. Palpi and head ochreous, former shaded with purple-brown; thorax largely purplish; abdomen ochreous, tufted with black; primaries longer and narrower than is usual in this genus, purple brown, suffused with white, which color is predominant in basal and costal half of median areas; basal area slightly tinged with ochreous; t. a. line broad, oblique, ochreous, preceded in lower portion by a deep purple-brown scale tuft and bordered outwardly with purple; discal dots almost obsolete, very small, separate; t. p. line very indistinct, indicated by slight blackish marks on veins; s. t. line wanting; fringes reddish purple. Secondaries pale smoky, hyaline, with darker border and pale ochreous line at base of fringes which are tinged with smoky. Beneath, primaries pale smoky with ochreous costa, black dot at base and short black streak in cell; secondaries whitish with ochreous costa and black streak on subcostal vein extending from base to middle of wing and faint black mark on median vein beyond base.

♀. Deeper purple than ♂ with less white suffusion; t. a. line tending to purplish; t. p. line slightly better defined, dentate on the veins, followed by similar s. t. line very close to outer margin, the two separated by a pale shade; a median oblique shade line above inner margin. Expanse 18 mm.

HABITAT. Everglade, Fla. (Apr. 8-15). Ft. Myers, Fla. (Apr. 16-23). 1 ♂, 2 ♀. Types, Coll. Barnes.

The markings on underside of ♂ would ally this species with *angusella* Grt.

IMMYRLA BUMELIELLA sp. nov. (Pl. IV, Fig. 7).

Thorax and primaries light gray; t. a. line crossing centre of wing purple-brown, more or less geminate, filled with pale gray, irregular, angled slightly inwardly in cell and below fold, preceded by a broad pale purple-brown shade extending from inner margin upwards to cell, the inner margin of this defined by a line of raised black scales; two prominent black superimposed spots at end of cell; from a point below these on inner margin a more or less prominent purplish shade extends obliquely to apex of wing; t. p. line purple-brown, angled strongly inwardly at cell, then rather straight and very slightly dentate to inner margin near anal angle, edged outwardly by a line of pale gray which in turn is followed by slight purplish suffusion; terminal portion of wing gray with dark terminal line, not attaining apex of wing; fringes smoky. Secondaries hyaline white, slightly smoky along terminal border with ochreous line at base of fringes. Beneath, primaries smoky, secondaries as above. Expanse 17 mm.

HABITAT. Ft. Myers, Fla. (Apr. 24-30) (May 1-7). 1 ♂, 5 ♀. Types, Coll. Barnes.

The species was bred from boat-shaped cocoons attached firmly to the twigs of a bush, which has been identified for us through the kindness of Mr. W. T. Davis as *Bumelia microcarpa*. We place the species in *Immyrla* Dyar rather than *Salebria* Zell. on account of the raised tuft of scales on primaries.

MESCINIA? ESTRELLA sp. nov. (Pl. III, Fig. 4).

♀. Palpi upturned to well above front, primaries 11 veined, 2 and 3 well stalked, 4 and 5 stalked, 6 from below upper angle, 8 and 9 stalked, 10 and 11 free; secondaries 7 veined, 2 from well before angle of cell, 3 and 5 from a point, 8 very short, on long stalk with 7; primaries purplish, pale brown along inner margin and broadly suffused with whitish along costa; t. a. line dark, most distinct in costal portion with outward angle in cell and less prominent one in submedian fold; dark discal dots separate; t. p. line pale, almost rigidly oblique from costa 1/4 from apex, preceded by dark shade line and edged outwardly in costal portion with fine dark line; dusky terminal line. Secondaries smoky, hyaline. Beneath, smoky, secondaries as above. Expanse 13 mm.

HABITAT. Everglade, Fla. (Apr. 8-15). 4 ♀. Type, Coll. Barnes.

Not having a ♂ we place the species temporarily in the genus *Mescinia Rag.* which is the only genus we know in which veins 2 and 3 and 4 and 5 of primaries are stalked and secondaries are 7 veined; our species differs from Ragonot's definition in having vein

10 free. We have been unable to find any description that would fit our species.

GENUS DIVITIACA gen. nov. (Type *Divitiaca ochrella* sp. nov.).

Palpi porrect, twice the length of head; maxillary palpi small, filiform; proboscis present; antennae of ♂ with small scale tuft on 2nd joint, ciliate, of ♀ simple; primaries 10 veined, 4 and 7 wanting, 2 from shortly before angle, 3 and 5 connate, from angle, 6 from below upper angle, 8 and 9 stalked, 10 separate, 11 well removed from 10; secondaries with cell less than 1/2 the length of wing, the lower angle produced, 7 veined, 4 wanting, 2 from angle, 3 and 5 stalked, 6 from upper angle, 7 and 8 stalked.

The genus would fall near *Diviana* Rag. differing in the long porrect palpi and scale tuft at base of antennae.

D. OCHRELLA sp. nov. (Pl. I, Fig. 3).

Head, thorax and primaries pale ochreous, veins on latter slightly paler; a small black patch above inner margin near base; t. a. line bent strongly outwardly, composed of blackish spots on the veins more or less conjoined; a single black dot on median vein at end of cell, t. p. line formed by black shading on the veins, subparallel to outer margin; slight terminal black spots in central portion of wing. Secondaries whitish, semi-hyaline, with slight smoky shading at apex. Beneath as above, but immaculate. Expanse 17 mm.

HABITAT. Everglade, Fla. (Apr. 8-15). 2 ♂, 3 ♀. Types, Coll. Barnes.

D. SIMULELLA sp. nov. (Pl. I, Fig. 6).

Smaller and deeper in color of primaries than the preceding species, more or less suffused with pinkish at base and above inner margin; maculation almost identical, dark basal shading rather more extended and terminal dark dots more prominent. Secondaries pale hyaline, smoky. Expanse 13 mm.

HABITAT. Everglade, Fla. (Apr. 8-15). 2 ♂, 2 ♀. Types, Coll. Barnes.

Should be readily separated from *ochrella* by smaller size, deeper color of primaries and smokier secondaries.

D. PARVULELLA sp. nov. (Pl. I, Fig. 9).

The smallest of these three closely allied species. Pale ochreous, costa rather broadly whitish; t. a. line represented by a rather prominent dark dot in the cell and some dark sprinkling of scales above inner margin; t. p. line distinct, black, slightly broken, parallel to outer margin; two minute discal spots surrounded by slight powderings of black scales; terminal black dots. Secondaries pale smoky, semi-hyaline. Expanse 11 mm.

HABITAT. Marco, Fla. (Apr. 16-23). 2 ♂. Type, Coll. Barnes.

Distinguished by the paler costa and rather more distinct t. p. line.

HOMOEOSOMA DIFFERTELLA sp. nov. (Pl. IV, Fig. 9).

Primaries rather pale even gray, shiny, slightly paler along costa to end of cell and sparsely sprinkled with smoky scales; t. a. line represented by a dot below costa and a larger one above inner margin; two superimposed discal dots, minute in ♂, larger in ♀; obscure blackish shading in position of t. p. line; secondaries hyaline smoky, slightly darker outwardly with white fringes. Beneath, primaries smoky, secondaries as above. Abdomen pale ochreous rather heavily shaded with smoky in ♂. Expanse 18 mm.

HABITAT. Everglade, Fla. (Apr. 8-15). 1 ♂, 1 ♀. Types, Coll. Barnes.

Paler and less heavily marked than *mucidella* Rag.

VARNERIA ATRIFASCIELLA sp. nov. (Pl. III, Fig. 7).

Primaries deep vinous red crossed by a broad blackish median band, broader at costa, enclosed between two pale ochreous lines. Secondaries smoky hyaline. Expanse 9 mm.

HABITAT. Everglade, Fla. (Apr. 8-15). 2 ♀. Type, Coll. Barnes.

Closely allied to *postremella* Dyar, of which it is possibly but a variety.

ANERASTIINAE

ALAMOSA BIPUNCTELLA sp. nov. (Pl. I, Fig. 7).

Primaries pale ochreous sprinkled sparsely with black along the veins which are slightly paler than the ground; a prominent round black spot at end of cell; slight dark terminal line above anal angle. Secondaries white with smoky terminal line. Beneath, primaries ochreous, secondaries white. Expanse 15 mm.

HABITAT. Ft. Myers, Fla. (Apr. 16-23). 1 ♀. Type, Coll. Barnes.

BANDERA CARNEELLA sp. nov. (Pl. I, Fig. 5).

Primaries pale flesh-color, sparsely sprinkled with black atoms; costa very slightly paler; two small black superimposed spots at end of cell. Secondaries and abdomen yellowish white. Expanse 11 mm.

HABITAT. Everglade, Fla. (Apr. 8-15). 1 ♂. Type, Coll. Barnes.

The venation agrees with the definition of *Bandera*, 3 and 5 of secondaries being merely rather shorter stemmed than usual. The palpi in our specimen are rather mutilated, but seem more as in *Tampa* Rag. than as in *Bandera;* for the present, however, we place the species under the latter heading.

PTEROPHORINAE

PLATYPTILIA CRENULATA sp. nov. (Pl. III, Fig. 8).

Primaries reddish fawn-color, lobes shaded with smoky and crossed by a distinct pale line 2/3 from base of incision; preceding this line on the upper lobe is a distinct black dash in central portion and dark costal shading; just beyond base of the incision is a large smoky triangular patch on costa extending downward slightly beyond incision; 1/2 way between base of wing and incision a black dot; a similar one on inner margin nearer base and a smaller dot at base of wing; costa slightly shaded with dusky, cut with ochreous; apex of upper lobe pale, inner angle of lobe 1 and upper angle of lobe 2 with slight black streak; outer margin distinctly crenulate, the bulging portions marked with black, the incisions ochreous; beyond this the fringe is pale, unicolorous, slightly dotted with black along inner margin. Secondaries smoky with apical black scale patch on 3rd lobe and scattered black scales along inner margin to base of wing. Beneath as above but smoky in color. Abdomen and legs pale ochreous, latter slightly blackish at base of spurs. Expanse 12-14 mm.

HABITAT. Ft. Myers, Fla. (May 1-7); Everglade, Fla. (Apr. 1-15); Chocoloskee, Fla. 11 specimens. Type, Coll. Barnes.

This species would fall, according to Fernald's table, near *acanthodactyla* Hbn. We found it fairly common along a roadside at Ft. Myers sitting on the stems of what Mr. Davis told us was a *Eupatorium* species, but which had frond-like leaves more like *Euphorbia*.

PTEROPHORUS UNICOLOR sp. nov. (Pl. I, Fig. 8).

Primaries pale straw-color, immaculate, slightly tinged with smoky along terminal margins of lobes; first lobe pointed, second lobe rather broad with well defined upper angle. Secondaries pale smoky with lighter silky fringes. Beneath, smoky. Legs pale ochreous, first two pairs blackish inwardly. Expanse 14 mm.

HABITAT. Marco, Fla. (Apr. 24-30). 1 ♀. Type, Coll. Barnes.

From several larvae, found boring in the stems of the *Eupatorium* species above referred to, we succeeded in breeding this single specimen.

PTEROPHORUS CERVINICOLOR sp. nov. (Pl. IV, Fig. 10).

Primaries with both lobes narrow, pointed and slightly downcurved at apex, deep grayish fawn-color, a slight sprinkling of blackish scales on inner margin 1/4 from base, forming a more or less obvious patch, a dark dot 1/2 way between base of wing and incision; two slight black costal dashes near apex of wing and another on inner margin of 1st lobe near apex; extreme apex of both lobes tipped with black and fringe along inner margin of 2nd lobe rather regularly cut by black; fringes otherwise pale fawn. Secondaries deep smoky with pale fawn fringes. Beneath, smoky, lobes of primaries paler. Expanse 16 mm.

HABITAT. Everglade, Fla. (Apr. 8-15). 2 specimens. Type, Coll. Barnes.

The color of this species is much as in the darker forms of *monodactylus* Linn. but the shape of the 2nd lobe would prevent it from being considered a small form of this species.

STENOPTILIA PALLISTRIGA sp. nov. (Pl. IV, Fig. 11).

Primaries bright fawn-color, deeper along costa and on first lobe, this latter with paler horizontal dash near apex, preceded by a few black scales; both lobes sprinkled with white scales; a black dot 1/2 way between base of wing and incision from which a sprinkling of black scales extends to a second black dot at base of incision; fringes of 1st lobe white with broken black line at base of 2nd lobe, dusky with black basal dots in apical portion of lobe. Secondaries deeper brown with paler silky fringes, 1st lobe with slight black point on inner margin near apex. Beneath much as above with the paler dash of first lobe more distinct. Legs whitish. Expanse 18 mm.

HABITAT. Ft. Myers, Fla. (May 1-7). 1 ♂. Type, Coll. Barnes.

PLATE I

All specimens x 2.

1. *Aresia parva.* Type ♂.
2. *Celama obliquata.* Type ♀.
3. *Divitiaca ochrella.* Type ♂.
4. *Microcausta flavipunctalis.* Type ♂.
5. *Bandera carneella.* Type ♂.
6. *Divitiaca simulella.* Type ♂.
7. *Alamosa bipunctella.* Type ♀:
8. *Pterophorus unicolor.* Type ♀.
9. *Divitiaca parvulella.* Type ♂.

PLATE I

PLATE II

1. *Anomis serrata.* Type ♂.
2. *P. scapha argentimacula.* Type ♂.
3. *L. albicerealis floridalis.* Type ♀.
4. *Storteria unicolor.* Type ♂.
5. *Tyrissa multilinea.* Type ♂.
6. *Jocara perseella.* Type ♂.
7. *Tetralopha querciella.* Type ♂.
8. *Tetralopha querciella.* Type ♀.
9. *Thaumatopsis floridella.* Type ♂.
10. *Sylepta masculinalis.* Type ♂.
11. *Platytes punctilineella.* Type ♂.
12. *Herculia sordidalis.* Type ♀.

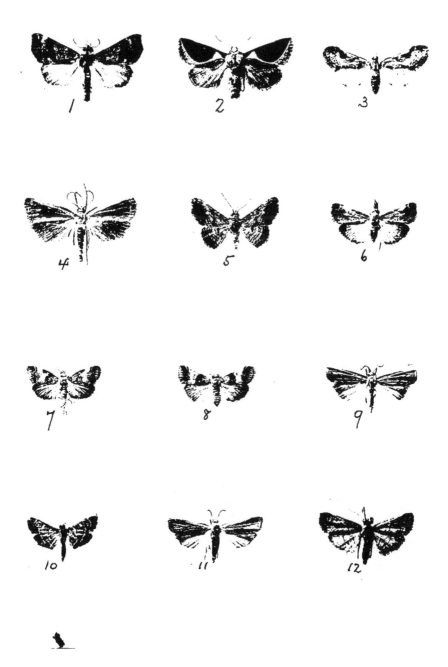

PLATE III

All specimens x 2.

1. *Acidaliodes eoides.* Type ♀.
2. *Patissa flavifascialis.* Type ♂.
3. *Davisia singularis.* Type ♂.
4. *Mescinia estrella.* Type ♀.
5. *Prochalia pygmaea.* Type ♂.
6. *Macrotheca unicoloralis.* Type ♂.
7. *Varneria atrifasciella.* Type ♀.
8. *Platyptilia crenulata.* Type ♂.

PLATE III

194

PLATE IV

1. *Piletocera simplicialis.* Type ♀.
2. *Nacoleia hampsoni.* Type ♀.
3. *Manatha nigrita.* Cotype ♂.
4. *Patissa sordidalis.* Type ♂.
5. *Proroblemma testa.* Type ♀.
6. *Stenoptycha solanalis.* Type ♀.
7. *Immyrla bumeliella.* Type ♂.
8. *Acrobasis tenuella.* Type ♂.
9. *Homoeosoma differtella.* Type ♂.
10. *Pterophorus cervinicolor.* Type.
11. *Stenoptilia pallistriga.* Type ♂.
12. *Glaphyria basiflavalis.* Type ♀.

1.	2.	3.

4	5	6.

7

8	9.

10	11

12

INDEX

	Page			Page
Aresia	167		ochrella	183
v. argentimacula	171		pallistriga	186
atrifasciella	184		parva	167
basiflavalis	172		parvulella	183
bipunctella	184		perseella	180
bumeliella	182		Prochalia	171
carneella	184		punctilineella	177
cervinicolor	185		pygmaea	171
crenulata	185		querceella	180
Davisia	179		serrata	169
differtella	184		simplicialis	175
Divitiaca	183		simulella	183
estrella	182		singularis	179
eoides	166		solanalis	174
flavifascialis	176		sordidalis (Herculia)	175
flavipunctalis	174		sordidalis (Patissa)	176
v. floridalis	173		Storteria	178
floridella	177		tenuella	181
hampsoni	173		testa	168
masculinalis	172		unicolor (Pterophorus)	185
multilinea	168		unicolor (Storteria)	178
nigrita	170		unicoloralis	176
obliquata	166			

TO THE

NATURAL HISTORY

OF THE

LEPIDOPTERA

OF

NORTH AMERICA

VOL. II
No. 5

SYNONYMIC NOTES ON
NORTH AMERICAN LEPIDOPTERA

BY

WILLIAM BARNES, S. B., M. D.

AND

J. H. McDUNNOUGH, Ph. D.

DECATUR, ILL.
THE REVIEW PRESS
AUGUST 6, 1914

Published
Under the Patronage
of
Miss Jessie D. Gillett
Elkhart, Ill.

INTRODUCTION

In the autumn of 1913 Dr. McDunnough visited Europe with a large collection of North American Lepidoptera for the purpose of comparing this material with the numerous types existant in the various European museums. The result of his visit to M. Charles Oberthur at Rennes has already been given to entomologists in Vol. IX of the "Études de Lépidoptérologie Comparée" published by Mr. Oberthur, in which a large number of Boisduval's types of Diurnals are figured; unfortunately Mr. Oberthur's time was very limited and no work could be done on the Noctuidae or Geometridae in his collection. In the National Museum at Paris Ragonot's types and the few remaining types of Guenée were examined; in Berlin a few of Zeller's oldest types and some of Moeschler's were discovered; the bulk of the work was done in the British Museum at London, where a month was spent in carefully going over Walker's, Grote's, and Zeller's types, with the exception of the Tortricids and Tineids. In all the above museums every facility for work was afforded us, and we wish here to express our thanks to the various curators and members of the museum staff who. combined to make our stay in Europe so pleasant and profitable.

The following notes give the main results of our comparisons; in the Noctuidae, thanks to the efforts of Sir Geo. Hampson and Mr. Wolley Dod, comparatively few mix-ups were noted; in several groups of Geometridae, viz., those not yet revised by Mr. Prout, great discrepancies were evident and some time must elapse before we can hope for anything like stability of nomenclature in this group; the Pyralids, with the exception of some few small groups, showed a distinct improvement over the Geometridae.

NOCTUIDAE

CHABUATA FISTULA Harv. (syn. *Tricholita ulamora* Sm.)

We can see nothing to warrant the supposition that *ulamora* Sm. from San Diego is a distinct species. See our remarks in "Contributions," Vol. I, No. 4, p. 55.

XYLOMANIA PERLUBENS Grt. (syn. *subapicalis* Sm.).

Smith was incorrect in stating that *perlubens* Grt. was the ♂ of *rubrica* Harv. (Jour. N. Y. Ent. Soc. XIX, 140) and that his *subapicalis* was distinct. We have specimens compared with both types and they are identical. *Rubrica* Harv. is a distinct species as listed. Our figure (Contributions II, pt. 1, Pl. 9, fig. 16) represents *perlubens*.

CIRPHIS SCIRPICOLA Gn. (syn. *Leucania calpota* Sm.)

One of Smith's co-types has been compared with Guenée's type and found identical.

ONCOCNEMIS CHANDLERI Grt. (syn. *colorado* Sm.)

In the type specimen of *chandleri* the orbicular and reniform are joined, but this feature is variable as shown by a long series in Coll. Barnes. In the other points of maculation we can see no difference between the two species.

PARASTICHTIS AURANTICOLOR Grt.

The type specimen in the British Museum, a ♂ labelled "Colorado," does not coincide with the remainder of the series nor with the figure given by Hampson (Cat. Lep. Phal. Pl. 109, fig. 30), which is not a figure of the type, but represents the usually accepted idea of *auranticolor*, an idea due probably to an erroneous determination by J. B. Smith in his revision of the genus Hadena; the same species is figured in our "Contributions" Vol. II, No. 1, Pl. XV, fig. 5-7. We note that Grote in his original description gives the locality as "Twin Lakes, Colo. Territory (Coll. Theo. Mead, No. 41)" whereas Hampson states "Colorado (Snow)" as the locality of the type specimens; our own note on the locality has been already given; there may just possibly have been a shifting of type labels, but in view of the fact that

Grote's meagre description would apply equally well to both species under discussion we are of the opinion that it would be well to regard the specimen so labelled in the British Museum as the authentic type. The true *auranticolor* may be separated from the spurious one by its shorter chunkier primaries and its rather brighter and more variegated appearance; the t. a. line is less dentate, the t. p. line is not *concave* in the fold but either straight or rather *convex* and edged inwardly with black at this point, and the s. t. space is *not* crossed between veins 1 and 2 by a dark shade. There is a long series in Coll. Barnes from Truckee, Calif., which agrees with *auranticolor* type; the locality is rather far removed, but there seems not much doubt as to the identity. *Auranticolor* is very close to *barnesi* Sm., the markings being practically identical and it is even possible that the original series included some *auranticolor*. In view of this fact we would restrict the name *barnesi* to those specimens from Yellowstone Park, Wyo., which are characterized by their general yellow-brown appearance; two co-types from this locality are in Coll. Barnes and one was figured in "Contributions" Vol. II, No. 1, Pl. 17, fig. 2. For the species commonly labelled *auranticolor* in collections and which is now apparently without a name, we would propose the name **GROTEI** and have labelled as type the ♀ specimen figured in our "Contributions" Vol. II, pt. 1, Pl. XV, fig. 7 from White Mts., Arizona. This species extends from Colorado southward through New Mexico and Arizona.

TRACHEA FUMEOLA Hamp. (syn. *T. probata* B. & McD.)

Hampson figures a California ♀ which is not the same species as the ♂ type from Arizona, this latter proving to be identical with what we described later, misled by the erroneous figure, as *probata*.

TRACHEA MODIOLA Grt. (syn. *Trachea mactatoides* B. & McD.)

It is probable that *Trachea mactatoides* B. & McD. will prove a synonym of this species; the type localities of the two species are respectively Wisconsin and Arizona but from what we have seen of other material of Grote's bearing the same locality label we have doubts as to the authenticity of this locality, (see our remarks under *Ozarba aeria* Grt.). Several species attributed by Grote to Racine, Wis. appear to have been originally collected in southern localities.

TRACHEA IMPULSA Gn.

We doubt if *mixta* Grt. should be placed as a synonym of this species; the differences as given by Hampson (Cat. Lep. Phal. VII, 210) appear to us to be of specific value.

AGROPERINA OBLIVIOSA Wlk.

This has been listed by Hampson as a synonym of *lateritia* Hbn.; we think however it may prove to be a good species; it is an even gray-brown, not *reddish,* without any contrasts, an indistinct clavi-form, a rather dentate t. p. line with slight white shading at inner margin and an obscure s. t. line without dark shading; the type locality is Rocky Mts. (Lord Derby).

EREMOBIA ALBERTINA Hamp.

This is apparently only a slight varietal form of *claudens* Wlk.; we can see nothing that would warrant its retention as a distinct species.

ACRONYCTA HAMAMELIS Gn.

The specimen in the British museum labelled in Guenée's hand-writing "hamamelis" and considered by Hampson a co-type is a worn specimen of *afflicta* Grt. We doubt if it can be held as a co-type, for Guenée states his specimens came from "Coll. Bdv." which is with Mr. Oberthur at Rennes; we regret not to have seen this species when paying our short visit to this noted collection.

ACRONYCTA LOBELIAE Gn.

The so-called type in the British Museum must be regarded as spurious since Guenée states the type specimen is in "Coll. Bdv." In any case it is *furcifera* Gn. of Smith's monograph and this is possibly the cause of Butler's rediscription under the name of *grotei* of what commonly goes as *lobeliae* Gn. If the types are in existence they will be with M. Oberthur at Rennes.

ARZAMA DENSA Wlk. (syn. *Sphida oecogenes* Dyar)

The type of *densa* Wlk. proves to have a frontal protuberance and is quite distinct generically from *Bellura gortynoides* Wlk. The genus *Arzama* Wlk. will therefore take precedence over *Sphida* Grt. There are several specimens of the type lot of *oecogenes* Dyar in Coll. Barnes, received years ago from Prof. Riley; one of these agreed exactly with the type of *densa* Wlk. As it now stands the synonymy of the genus is as follows:

ARZAMA Wlk. (*Sphida* Grt.)

>   obliqua Wlk. (syn. *obliquata* G. & R.)
>   densa Wlk. (syn. *oecogenes* Dyar)
>   gargantua Dyar.
>   anoa Dyar.

OZARBA AERIA Grt.

The species reported by us as *O. fannia* Druce (Can. Ent. 45, 184) proves to be *aeria* Grt.; *fannia* must therefore be removed from our N. American lists; the locality "Wisconsin" as given by Grote is probably an error as our series of specimens comes from Southern Texas; Dr. Hoy of Racine, Wis., from whom Grote received various specimens, does not appear to have attached much importance to locality labels, and Grote has simply attributed all the material to Wisconsin.

EUSTROTIA ORTHOZONA Hamp. (syn. *E. santa rita* Dyar)

Dr. Dyar has misidentified Hampson's species, evidently considering it to be the form with a tooth about the centre of the inner edge of the black median band. As a matter of fact this edge in both types is almost straight; the toothed form is probably the more normal form of *antonita* Dyar, which was described from a single specimen received from Dr. Barnes in which the median band is broken. Dr. Dyar would do well to follow his own advice to us and avail himself of the resources of the various European and American museums before appearing in print with needless synonyms.

GRAEPERIA NUICOLA Sm. (syn. *G. carcharodonta* Hamp.)

In his revision Smith appears to have overlooked the peculiar frontal prominence of both this species and *meskei* Sm., this latter species being apparently merely a slight variety of *nuicola;* the types of both species in the National Museum show this structure, *nuicola* being identical with the species described later by Hampson as *carcharodonta.*

GRAEPERIA ALTERA Sm. (syn. *conocharodes* Hamp.)

Smith's species had been misidentified in the British Museum thus leading to the above synonymy.

HOMOPYRALIS TANTILLUS Grt. (syn. *H. monodia* Dyar).

Another misidentification on the part of Dr. Dyar which could have been obviated by sending material to the British Museum for comparison. *Tantillus* of the National Museum is a species allied to *quadrisignata* Wlk. and nothing like Grote's species; we have specimens which have been compared with both of the above types and find them identical.

ZANCLOGNATHA JACCHUSALIS Wlk. (syn. *marcidilinea* Grt.)

Walker's species has been sunk as a synonym of *cruralis* Gn. Whilst the type is a very poor specimen it still distinctly shows a slightly dentate t. a. line, which would at once separate it from Guenée's species and make Grote's species a synonym.

RENIA FACTIOSALIS Wlk. (syn. *clitosalis* Wlk.)

The latter type is very poor, but shows a distinct brownish shade to primaries and should we think become a synonym of *factiosalis;* a second specimen associated with the type in the collection is certainly this latter species. *Tilosalis* Sm. is very close to this species and we imagine it may prove identical although we hesitate at present to sink it in the synonymy.

ANOMIS SERRATA B. & McD.

Dr. Dyar, in his review of our "Contributions, Vol. II, No. 4," made this a synonym of *xanthindyma* Bdv. Sir Geo. Hampson however stated to us that *xanthindyma* Bdv. described from Madagascar, is *not* the form with serrate antennae, backing his assertion with specimens from the type locality. We do not know on what Dr. Dyar has based his statement, but it would almost seem as if our name *would* hold after all, in spite of the veto from Washington.

NOTODONTIDAE

HETEROCAMPA ASTARTE Dbldy. (syn *chapmani* Grt.)

A ♂ specimen in the British Museum marked "E. D. St. John's Bluff, E. Fla.", a locality corresponding with that of the original description, can, we think, without doubt be regarded as the authentic type. The ♀ type of *chapmani* Grt., also from Florida, is the same species; we can see no resemblance to Felder's figure of *picta* of which it is at present listed as a synonym, nor can we agree with Mr. Schaus' statement regarding these species (Trans. Ent. Soc. Lond. 1901, p. 302, 303). This species is distinguished by its greenish ash color with prominent white lunule curving downward from near apex and bordered outwardly with *red-brown;* there are also red-brown terminal lunules. Packard's colored figure in the monograph (Pl. VII, fig. *27*) is not typical; it represents possibly the Texan variety.

HETEROCAMPA VARIA Wlk. (syn. *H. obliqua* Pack.)

Walker's type ♀ is distinct from *astarte* Dbldy. and is the same species as *obliqua* Pack. The Monograph gives this synonymy but Packard has adopted his own name which has been followed by later authors.

LIPARIDAE

HEMEROCAMPA PLAGIATA Wlk. (syn. *H. definita* Pack.)

We had listed this much misidentified species, on the strength of a photograph of the type, as a synonym of *H. leucostigma* A. & S. (Contrib. Vol. II, No. 2, p. 50). A personal examination of the type shows that it is not that species, but *definita* Pack., and the name will therefore have to be resurrected.

GEOMETRIDAE

EUCYMATOGE ANTICARIA Wlk. (syn. *implicata* Wlk., *explanata* Wlk.)

The types of *implicata* and *explanata* are almost identical as far as we could judge; *anticaria* is a slightly better marked form, but not specifically distinct in our opinion.

VENUSIA CONDENSATA Wlk.

This is a synonym of *Euchoeca lucata* Gn. and not of *comptaria* Wlk. as listed in Dyar. We could not find the type of *inclinataria* Wlk. also listed as a synonym.

HYDRIOMENA FRIGIDATA Wlk.

Dr. W. T. M. Forbes who saw a specimen we had compared with the type of this species was of the opinion that *transfigurata* Swett would prove a synonym; certainly Swett's description would fit our compared specimen well. *Renunciata* Wlk. seemed to us to be a rather grayer and uncontrasted form of the same species, but we could match it with nothing in our material.

HYDRIOMENA BISTRIOLATA Zell.

The type ♀ from Dallas, Texas, is quite a different species to that from Colorado identified for us as *bistriolata* by Mr. Swett. It is a long palpi form with distinct greenish shade to the paler areas; the black band is not so oblique nor is the outer edge so irregular as in the Colorado specimens. We have nothing to match it in Coll. Barnes.

HYDRIOMENA TAENIATA Steph.

This name should be dropped from our lists; it applies to European specimens; *explagiata* Wlk. should be used in its place. *Basaliata* Wlk. (syn. *Eucymatoge grandis* Hlst.) may prove to be a good species; a series in Coll. Barnes from the west coast of British Columbia is larger with the markings less contrasted.

TRIPHOSA PROGRESSATA Wlk.

This should be omitted from our lists; the type from S. Domingo is a *Scotosia* and in the British Museum is sunk as a synonym of *affirmata* Gn.; *dubitata* L. should also be dropped as applying only to European specimens. This will leave *indubitata* Grt. (type loc. Montreal, Que.) applicable to our Eastern specimens; the larger form from the Pacific coast, if distinct from the Eastern race, is without a name in its normal form; the name *pustularia* Edw., given to the peculiar mottled form, could however be applied to the whole species if the eastern and western races be found distinct specifically. This mottled form also occurs in Eastern specimens.

PETROPHORA CONGREGATA Wlk.

This species, at present sunk in the synonymy of *unangulata* Haw. by some utter misidentification of Hulst's is the one commonly called *abrasaria* H. S. If necessary therefore this name may be used in preference to *abrasaria* which typically is European.

EUSTROMA PROPULSATA Wlk.

The synonymy as given in Dyar's list is very incorrect, *propulsata* is the yellow species, formerly listed as *populata* L. with type of maculation similar to *luteolata* Hlst.; it occurs all through Canada from Newfoundland (type locality) to Vanc. Is. *Molliculata* Wlk. is a very good species, ranging in color from deep yellow to brown, with red-brown subapical patch preceded by white line; we have typical specimens from New Brighton, Pa. *Remotata* Wlk. is another good species very close to, if not identical with, *destinata* Moeschler. *Similis* Wlk. is a species of which *harveyata* Tayl. may be a synonym; the two are very closely related, *schistacea* Warr. forming a third member of the group; we had nothing with us that would exactly match Walker's type but from our notes the type of *harveyata* in Coll. Barnes would approach it very closely. This species is yellowish with broad purple basal and median bands; it is rather rare but apparently wide spread in various forms through the north and the Rocky Mts. *Cervinifascia* is a *Mesoleuca* with which we had no material to match but of which *mirandata* Tayl. may prove to be a synonym; the proximity of the type localities would rather point to this conclusion and Taylor's type before us fits in with our brief notes. However further comparison is necessary before definite statements can be made.

EUSTROMA CUNIGERATA Wlk.

This is the form of *explanata* Wlk. in which the median dark band breaks up in the central portion into oblong oval spots.

MESOLEUCA INTERMEDIATA Gn.

The type specimen is at Paris and is the species commonly known in N. American collections as *lacustrata* Gn., i. e., the form with the *white* subterminal area and *large discal spot,* both of which features are mentioned in the original description; the two names appear to have been interchanged but we did not see the type of *lacustrata* to make certain; *ideata* Wlk. (nec *iduata* Gn.) is also this species; it is at present reposing in the synonymy of *fluctuata* L. A specimen of this species was probably responsible for the listing of *unangulata* Haw. from N. America which name we consider should be dropped from the lists.

FIDONIA ATOMARIA Gn.

The type of this species (Sp. Gén. X, 155) is at Paris and bears a label by Mr. Th. Mieg stating that the type locality New Holland as given by Guenée is incorrect and that the species is N. American and the same as *faxoni* Minot, over which it has priority. As far as we could judge this is correct although we had no actual specimens for comparison.

EUFIDONIA FIDONIATA Wlk.

The type is a ♂ with very heavy subterminal and terminal bands on primaries leaving a white wavy s. t. line; *notataria* Wlk. is a ♀, white, speckled heavily with light brown, with less distinct banding on primaries than the preceding, but heavier markings on secondaries; this may be merely a sexual difference. *Discospilata* Wlk. seems to be a northern form; we have 3 ♀'s from Canada corresponding with the type; they are white with the brown markings much deeper black-brown than in *notataria* although the style of maculation is similar. In a variable species like this breeding is necessary to determine the specific standing of the forms.

MELLILLA INEXTRICATA Wlk.

The synonymy of the genus *Mellilla,* as it stands in our lists, is much muddled. The type of *inextricata* Wlk. from Florida is worn but the secondaries are pale orange *blotched with purple-brown* forming two semi-parallel bands across wing; we have seen two specimens like it. The species commonly known under this name is *xanthometata* Wlk. with *chamaechrysaria* Grt. as a synonym; both ♂ types are poor, but show distinctly the straight post median line followed over the remainder of the wing by a dark shade containing black blotches. From the description *rilevaria* Pack. would also be a synonym, merely differing from typical specimens in that the subterminal blotches are wanting. We have bred quite a series of this species from larvae on honey-locust and find this feature variable. The name *snoviaria* Pack. applies we think to the summer form of the same species in which the heavy subterminal dark shading is almost absent and the black blotches show more clearly; the spring ♂'s, and to a lesser degree the ♀'s, are very contrasted in this respect. The species would therefore stand as follows:

MELLILLA Grt.

> inextricata Wlk.
> xanthometata Wlk.
>> *chamaechrysaria* Grt.
>> *rilevaria* Pack.
> f. aest. snoviaria Pack.

SCIAGRAPHIA GRANITATA Gn.

The synonymy of this species as given in Dyar's list is appallingly muddled and our series of specimens is not nearly sufficient to attempt to do more than merely indicate certain of the gravest errors. The type specimen is at Paris and agrees well with Guenée's description; the lines are brownish on a white ground and the most marked feature is a brown spot on costa near apex. The following names at present included in the synonymy apply to good species and have nothing to do with *granitata:*

EXAUSPICATA Wlk.

This species falls into the *inquinaria-subfalcata* group of *Cymatophora* (of Dyar's list) and resembles considerably *julia* Hulst in general style of maculation; the type ♀ we have matched with a ♀ from New Brighton, Pa., and there are 5 ♂'s in Coll. Barnes from the same locality.

ORDINATA Wlk.

Very close to *Macaria eremiata* Gn. but larger, paler, with the 3 lines of primaries and 2 of secondaries distinct; we have matched this with a specimen from Chicago, Ill.

ABRUPTATA Wlk.

Our note on the type is that it is a round-winged species with dark spot on post-median line followed by a second spot subterminally and has nothing to do with *granitata* Gn.

UNIMODARIA Morr.

♂ type in the British Museum (ex Grote Coll.) is small, dark and unicolorous. We saw a specimen amongst material sent to Mr. Prout for determination by the late J. A. Grossbeck.

SUCCOSATA Zell.

Type has a brown costal patch near apex and strongly waved transverse lines much as in *proxanthata* Wlk.; Mr. Grossbeck's material contained a specimen of this species also; this may approach the true *granitata* more closely than some of the following forms.

The remainder of Walker's types, as far as we could find them, are more or less closely related and we are in doubt as to whether they will prove species or varieties. *Irregulata* Wlk. we have matched with a specimen from Biddeford, Me. It is a rather even gray form with single brown spot subterminally between veins 3 and 4; *quadrisignata* Wlk. is very similar but grayer and more faintly marked; *dispuncta* Wlk., *haliata* Wlk. and *fissinotata* Wlk. are also even gray forms with the brown spot split into two; *retinotata* Wlk. we have matched fairly well with specimens from New Brighton, Pa.; it is whiter in general appearance with stronger lines and brownish costo-apical spot. *Exnotata* Wlk. is very similar but larger; *submarmorata* Wlk. has no subterminal spot and is sprinkled above and below with whitish atoms.

We did not see the types of *contemptata* Gn. nor *haliata* Gn., they are doubtless with Mr. Oberthur and figures should be published in due course in his "Études"; *inordinaria* Wlk., *retractaria* Wlk. and *subapicaria* Wlk. are also unknown to us.

SCIAGRAPHIA HELIOTHIDATA Gn.

The species figured by us in our Contributions, Vol. II, No. 3, Pl. 6, fig. 1-3 and doubtfully referred to *heliothidata* should be placed under *nigrocomina* Warr.; *restorata* Wlk. and *subcolumbata* Wlk. refer to one species which is different to *nigrocomina* and may be the true *heliothidata* Gn. which type we did not see; in any case this W. Indian form will hardly occur in our fauna. A type of *duplicata* Pack. is in the British Museum and is referable to *ocellinata* Gn. which we have already pointed out to be a good species (Contributions, Vol. II, No. 3, Pl. 6, fig. 4-6).

SCIAGRAPHIA CONTINUATA Wlk.

The type from E. Fla. is poor but the postmedial line shows no angle below costa; *strigularia* Wlk. has this line angled and we have matched the type with a specimen from Black Jack Springs, Texas; *orillata* Wlk. is distinct but Mr. Grossbeck was mistaken in stating that it is a western species (Ent. News, XX, 352); it was described from Orillia, Ont., and we have matched the type exactly with a specimen from Hymers, Ont.; it very closely resembles the western form and we doubt if the two can be separated.

CYMATOPHORA BITACTATA Wlk.

. We think *atrosignata* Wlk. will prove a synonym although the two types do not exactly match; in *bitactata* the discal spot is distinct from the median band, whilst in *atrosignata* the two are joined forming a broad angle in the band; this latter form we have matched with a specimen from Hymers, Ont. Breeding will be necessary to settle the relationships of the various members of the *wauaria* group.

CYMATOPHORA VARADARIA Wlk. (syn. *C. abbreviata* Wlk.)

This species, at present lost in the synonymy of *Caberodes confusaria* Hbn., proved to be identical with the species known as *abbreviata* Wlk.; the types were widely separated in the museum collection, but on calling Mr. Prout's attention to the matter he carefully studied both types and confirmed our opinion.

APAECASIA DEFLUATA Wlk.

This name will have to be dropped from our lists; it was described from a specimen with unknown type locality and the type proves identical with that of *Tacparia zalissaria* Wlk. (Cat. Lep. Het. XX, 234) from Australia. This will leave the name *atropunctata* Pack. (*fernaldi* Grt.) available for our N. American species.

CARIPETA ANGUSTIORATA Wlk. (syn. *criminosa* Swett).

We were much surprised to find that that the type of *angustiorata* was the species figured by Packard (Monog. Pl. IX, fig. 52) in spite of Mr. Swett's article to the contrary (Jour. N. Y. Soc. XIV, 128). Apparently there was some error in comparing the types in the British Museum with Mr. Swett's specimens for the type seen by us was certainly authentic, agreeing in all respects with the original description. The only distinction between Walker's type and Packard's figure is that in the type the t. a. and t. p. lines actually touch, this being however a very variable feature. *Criminosa* Swett thus becomes a synonym of *angustiorata* Wlk. and *piniata* Pack. (*seductaria* Stkr.) will again be reinstated for the species with strigate primaries. *Latiorata* Wlk. from Florida is a good species and almost twice the size of *angustiorata*.

NEPYTIA SEMICLUSARIA Wlk. (syn. *fumosaria* Stkr.)

Specimens compared with both types are identical, both referring to the large suffused gray form. For the· smaller form, common in the Atlantic States, *canosaria* Wlk. is the correct name. Packard's figure (Monog. Pl. XI, fig. 32) is an excellent representation of *semiclusaria*. Pearsall in his paper on these species (Can. Ent. 39, p. 171) has misidentified Walker's species; it is *canosaria*, not *semiclusaria*, that has the yellowish front; if Pearsall be correct in his other remarks, then *pellucidaria* Pack. will become a synonym of *semiclusaria*, but we have not seen Packard's type so made no definite statements.

COENOCHARIS INFUMATARIA Grt. (syn. *ignavaria* Pears.)

This species, listed as a synonym of *Exelis pyrolaria* Gn., has been totally misplaced; a careful reading of Grote's description would have prevented such an error as he states that the size is 30 mm. and the build "stouter than *robiginosaria*" (*scolopacinaria* Gn.). An examination of the single ♀ type showed it to be a *Coenocharis* and we noted at the time that it must be close to *ignavaria* Pears. On returning home we sent a ♀ of this latter species to Mr. Prout who kindly compared it with the type and concluded that both represented the same species; he noted a few minor details of difference in the depth of incurve of the t. p. line in fold and the dark margin to wings on underside, but these are variable in our series and we think the two names represent but one species. Our identification of Pearsall's species was made by Mr. Grossbeck who had seen the type.

SELIDOSEMA HUMARIUM Gn.

The mass of names in the synonymy of this species according to our present lists is only surpassed by that of *granitata* Gn. and the muddle is equally great. Not having seen Guenée's types we can only doubtfully place the species but we follow the British Museum temporarily in our identifications of Guenée's species. *Humarium* is the small purplish brown species with t. p. line of primaries twice well incurved, once below cell and again in the fold and same line almost straight on secondaries; *illaudata* Wlk. and *intractaria* Wlk. are synonyms; the latter type ♂ is very worn, but we think it represents the same species.

Eriosata Wlk. is a peculiar pale form with strong dark lines referred in the museum collection to *pampinaria* Gn.

Expressaria Wlk. and *ephyraria* Wlk. are referred in the collection to *intraria* Gn.; this is quite distinct from *humarium* being a small rather mottled species with distinct ringlets and an angle to t. p. line of secondaries below costa; *takenaria* Pears. will come very close to this species.

Defectaria Gn. with *albigenaria* Wlk. as synonym is a larger whiter species, common to Florida and Texas, with a distinct red-brown band following t. p. line in ♂ and preceding t. a. line; the ♀ is without the banding and more speckled; we bred a series of this ex ova and the larval history was to have been incorporated in a list of Florida lepidoptera in course of preparation by Mr. Grossbeck; his untimely decease may delay the publication.

Transfixaria Wlk. is made a synonym of *sublunaria* Gn. which is separated from *pampinaria* Gn. on the strength of distinct discal ringlets and an angle in t. p. line of secondaries; possibly *areaxaria* Broadwell will fall here but we do not know the species for certain. We did not see the types of *momaria* Wlk. and *ejectaria* Wlk.

Melanolophia imitata Wlk.

The type ♂ from Vanc. Is. is identical with the species known as *subgenericata* Dyar.

Aethaloptera intertexta Wlk.

This species, listed in Dyar as *intextata,* was described from an unknown locality; it is not the same species as that commonly going under this name, being heavier marked and with stronger bend in t. p. line of secondaries; we could not match it and think for the present that the name *anticaria* Wlk. (*submuraria* Wlk.) should be used for our N. American species.

Therina scitata Wlk. (syn. *invexata* Wlk.)

We consider this a good species; it is at present in the synonymy of *fervidaria* Hbn., but differs in wing shape, the wings in ♂ being quite rounded and in ♀ showing only a very faint angle, much as in *pellucidaria* G. & R.; the t. p. line is also not angled, being only very slightly incurved in fold on primaries. The type ♂ of *scitata* is from Florida and type ♀ *invexata,* locality unknown; we have a pair from Florida, labelled by Dr. Dyar *fiscellaria* Gn., which agree with Walker's types.

Gonodontis duaria Gn.

Amyrisaria Wlk. is a synonym of this species, not of *Caberodes confusaria* as listed at present.

Euchlaena obtusaria Hbn.

Decisaria Wlk. *incisaria* Wlk. (Cat. Lep. Het. 35, p. 1546) and *muzaria* Wlk. all refer to this same species; the latter name is at present listed as a variety of *effectaria* Wlk., but only differs from *decisaria* in having the basal portion slightly yellower. The type of *effectaria* is not in the British Museum.

Euchlaena astylusaria Wlk.

Madusaria Wlk. is correctly placed as a synonym, Packard's fig. (Monog. Pl. XII, fig. 14) is correct; *deplanaria* Wlk. ♂ type is a small pale form without terminal shading, referable, we think, to this species, certainly not *amoenaria* Gn. as figured by Packard. *Oponearia* Wlk. and *tiviaria* Wlk. both refer to the same species, the former name having priority; this is a brown suffused form from E. Florida, which we could not match and which may be a distinct species.

Euchlaena deductaria Wlk.

Packard's figure (Pl. 12, fig. 11) represents this form fairly well although the type is rather more shaded outwardly with brownish; this is evidently the *pectinaria* of Guenée's description, but we have not seen the original description nor any of the figures mentioned by Guenée so cannot vouch for the correctness; Packard's figure (Pl. XII, fig. 18) is a ♂, not ♀, and represents another form also commonly going under the name of *pectinaria* D. & S.; this is *sirenaria* Stkr.; *propriaria* Wlk. is this species, a rather aberrational form in which the yellow has been largely replaced by brown sprinkling. If Guenée's determination be correct and *deductaria* Wlk. becomes a synonym of *pectinaria* Schiff, then *propriaria* Wlk. with yellow variety *sirenaria* Stkr. will apply to the larger species.

Eutrapela alciphearia Wlk.

Packard's figure (Pl. XII, fig. 27) represents this species well; it is the form with the t. a. line of primaries nearly straight and the underside of secondaries rather evenly dark beyond the lunule with the exception of a straight whitish postmedian narrow band; *perangulata* Hulst, according to a photograph of the type in our possession.

will fall into the synonymy. We have this species from Ont., Alberta and Vancouver Island. The summer form of this species, as is usual in the genus, is much smaller, brighter yellow on the upper side and with considerable pink and orange-yellow suffusion on the under side; we have a series of both sexes from Vancouver Is., B. C., and would distinguish it under the name *form. aest.* **ORNATA.**

Kentaria Grt. is a good species in our estimation; Packard's figure (Pl. XII, fig. 28) represents a small specimen, possibly the summer form. It appears commoner in the East than the preceding species; we have a series from New Brighton, Pa., and Cartwright, Man., but all are of early date although it is probable the species is two-brooded. In this species the t. a. line of primaries is much outcurved, the t. p. line usually bent inwards opposite the cell and the underside of secondaries brighter orange at the base and paler outwardly than in *alciphearia* Wlk.

SABULODES ARCASARIA Wlk.

Sulphurata Pack. falls as a synonym of this species. The ♀ stands in the British Museum under *constricta* Warr. but we cannot find that this name has been published.

LIMACODIDAE

EUCLEA INCISA Harv.

Dr. Dyar was in error in listing *incisa* Harv. as a synonym of *paenulata* Clem.; it is *excisa* Wlk. (Cat. Lep. Het. XXXII, 484), a name ignored by Dr. Dyar, that is the synonym. *Incisa* Harv. is quite a distinct species and the older identifications were correct; the name must be reinstated with *mira* Dyar as a synonym.

PYRALIDAE
PYRAUSTINAE

CONCHYLODES OVULALIS Gn.

This species is distinct from *platinalis* Gn.; we saw both types in Coll. Oberthur; our common species with even black subterminal line on primaries is *ovulalis; platinalis* has the apical and central portions of this line broadened into black blotches; the locality Missouri **may** be incorrect as we have seen no N. American material agreeing with the type.

LYGROPIA RIVUALIS Hamp. (syn. *Bleph. nymphulalis* Haim.)

Although we have not seen Haimbach's type, his description leaves no doubt as to the correctness of the above synonymy.

POLYGRAMMODES HIRTALIS Gn.

Capitalis Grt., listed in the synonymy, is a good species; it is much larger, whiter in color, with the s. t. brown line much more irregular and excurved; the type ♀ is from Florida, but we have a long series from Decatur, Ill. A much smaller, yellower form we have in numbers from Brownsville, Texas; this we take to be *hirtalis* Gn. although we have not seen the type; it is certainly *amatalis* Wlk.

LOXOSTEGE OPHIONALIS Wlk.

We can see nothing to separate *nasonialis* Zell. from this species.

GYROS MUIRI Hy. Edw. (*Monocona rubralis* Warr.)

Warren's species and genus becomes a synonym of Henry Edwards', originally described as a Noctuid, and still so listed by Dyar.

AUTOCOSMIA CONCINNA Warr.

This species must be separated from *nexalis* Hulst; while both species are practically identical in maculation, *concinna* is much smaller and bright red-brown at base and along costa of primaries, *nexalis* being generally more dingy brown; we have *nexalis* from the type locality in Coll. Barnes and *concinna* from Kerrville, Texas, both compared with the types.

CYBALOMIA EXTORRIS Warr.

This name has two years' priority over *Metasia quadristrigalis* Fern.; we have specimens of the latter species identified for us by Fernald himself.

PHLYCTAENIA ITYSALIS Wlk.

This species, together with *variegata* Wlk. and *turmalis* Grt. all represent the form with solid black reniform and orbicular and broad dark costal stripe; if any reliance can be placed on Dr. Dyar's description of *tillialis* (Proc. U. S. N. Mus. XXVII, 916) then this name will also apply to the same form; Dr. Dyar however is very careful not to state any points of distinction between his new species and *itysalis* Wlk., so, as we have not seen the type, we refrain from following his plan of at once listing it definitely as a synonym; there may be points of difference not brought out in the description.

PYRAUSTA MATRONALIS Grt.

This species should be transferred into the synonymy of *subsequalis* Gn.; the type, a ♂, is identical with specimens bred by us from thistle in this locality; *borealis* Pack. is a much smaller species with no marked sex distinction as in *subsequalis*.

NOCTUELIA SIMPLEX Warr.

This becomes a synonym of *Loxostege succandidalis* Hulst.

STENOPTYCHA SOLANALIS B. & McD.

This species is *not* synonymous with *pterophoralis* Wlk. as lately stated by Dyar; the type ♀ of this latter species has almost *hyaline* secondaries, is paler brown and lacks the dark costal dots of *solanalis;* the British Museum series is mixed. Again one of our descriptions has been justified and again the fallacy of rashly consigning names to the synonymy without an exact knowledge of the type specimens is proved.

NYMPHULINAE

ELOPHILA CLAUDIALIS Wlk.

The type has the black marginal dots of secondaries on a distinctly white ground with *no* black speckles; the form with black sprinkling is *medicinalis* Grt. and the two species are separated in the British Museum.

ELOPHILA CONFUSALIS Wlk.

This we consider distinct from *fulicalis* Clem. One of the best marks of distinction may be found on the secondaries; in *fulicalis* the oblique brown sub-basal band is followed outwardly by a clear white space which in turn is bordered by a *distinct brown line* forming a right angle a little inwards from the first marginal dot, counting from anal angle; in *confusalis* this distinct line is not present; there are various slight points of distinction on primaries, and *fulicalis* is rather larger. In Coll. Barnes are four specimens of *fulicalis* and a long series of *confusalis* from New Brighton, Pa.

CHRYSAUGINAE

CHALINITIS OBLIQUATA Hy. Edw. (syn. *albistrigalis* B. & McD.)

Dr. W. T. Forbes has been kind enough to send us a note on the venation of the species described by Hy. Edwards as *Earias obliquata* (Ent. Am. II, 9) drawn up from the type in the American Museum of Natural History. It turns out that this species, which has been omitted from Dyar's list, is a Chrysaugid, identical, without much doubt, with our species as listed above. It is satisfactory to have located one more of those "doubtfuls and unknowns", which are the bane of the entomologist's life.

SCHOENOBIINAE

RUPELA NIVEA Wlk.

We examined the venation of the type and found that on primaries veins 8, 9, and 10 were stalked whilst 11 became coincident with 12; the type is a rather small specimen measuring about 25 mm. in expanse; we exactly matched it with a ♀ from Everglade, Fla. Other specimens in Coll. Barnes are considerably larger and have vein 11 free, which leads us to wonder whether or not there may be two species associated here under one name. Our series is not however large enough to determine this. Our genus *Storteria* (type *unicolor* B. & McD.) differs from *Rupela* in having vein 11 of primaries free, a feature constant in 3 ♂'s, in Coll. Barnes; the peculiar long thoracic hairs are also absent and the cubitus of secondaries is distinctly haired, a fact which led us to place the genus in the *Crambinae;* as we already have stated (Can. Ent. 46, 31, 1914) it appears however much better situated in the *Schoenobiinae.* For the present it would be well not to sink *Storteria* as a synonym of *Rupela* until further studies have been made in the group.

SCIRPOPHAGA REPUGNATALIS Wlk. (syn. *consortalis* Dyar)

This species, described from unknown locality (1863 Cat. Lep. Het. XXVII, 144), proves the same as the species described by Dyar (1909 Proc. Wash. Ent. Soc. XI, 28) as *Argyria consortalis* from Florida; we have one of Dyar's types, ex Coll. Merrick, before us and several other Florida specimens. How Dr. Dyar came to make such an utterly erroneous generic reference is a riddle to us; in a very fresh ♂ the cubitus of hind wings shows no traces of hairs and the species bears great resemblance to *perstrialis* Hbn., merely lacking the white streak of primaries; we should not be surprised if it proved to be only a suffused form of Hübner's species, but of course breeding will be necessary to settle this question.

CRAMBINAE

EOREUMA DENSELLUS Zell.

In the original description Zeller states that he had had 5 poor specimens of this species in his collection from Texas for years unnamed and proposes *densellus* for them. In the British Museum are 5 specimens labelled *densellus* in Zeller's hand with further note "Texas, Stt. 68" the latter number evidently referring to the year; these we think are without doubt the original type specimens; they are all rubbed, just as stated by Zeller, but we have matched them with a ♂ in Coll. Barnes from Kerrville, Texas. It is a small species, quite narrow-winged with the veins marked in white; the venation is as given by Ely (Proc. Ent. Soc., Wash. XII, 204).

EPIPASCHIINAE

JOCARA INCRUSTALIS Hulst.

The species *perseella* B. & McD. from Florida has been summarily placed in the synonymy of Hulst's species by our ever-ready friend, Dr. Dyar. In the Hulst Collection at New Brunswick there are two specimens under this name, the one, a ♀, marked type, with no locality label, the other, very worn, labelled "Col." Neither bears much resemblance to our ♂ type of *perseella* and it would need to be a very variable species indeed for the two names to be considered as synonyms. We see no grounds therefore for accepting Dr. Dyar's dictum as correct.

TETRALOPHA MILITELLA Zell.

The types of this species, a ♂ and ♀ labelled "Carolina, Zimmermann", are in the Berlin Museum and were carefully examined. We have matched them exactly with specimens from Oconee, Ill., and are very much inclined to agree with Grote (Bull. U. S. Geol. Sur. IV, 691) that *Lanthape platanella* Clem. will prove a synonym. Unfortunately one has only Clemens' description to go by and this is obscure in several particulars; breeding from larvae on sycamore would of course settle the matter. Dr. Dyar (Proc. Wash. Ent. Soc. VII, 31) has made a pretty mess of the synonymy of this species and shown such a woeful lack of knowledge concerning the true species as to shame the veriest tyro in entomology. He lists as synonyms the following 4 species—*asperatella* Clem., *expandens* Wlk., *taleolalis* Hlst. and *fuscolotella* Rag. *Asperatella* Clem. with *expandens* Wlk. as a synonym is a totally different species, almost twice the size and quite differently marked; Dyar's ab. *clemensalis* of this species, with ochreous base of primaries, is in the Hulst Coll. marked *"nephelotella* Hlst. Type"; as however the locality label on this so-called type is "Blanco Co., Texas", and the original description calls for 'Penn." we have doubts as to the authenticity; further study is necessary.

Taleolalis Hlst. is a synonym of *subcanalis* Wlk. as far as it is possible to identify Walker's type specimen. We will make further remarks on these species later. *Fuscolotella* Rag. is a very good Arizona species; we have a series agreeing with Ragonot's type at Paris, a single ♂ marked "Ariz. (Morrison)"; it is smaller and blacker than *asperatella* with shorter palpi and approaches closest to *tiltella* Hulst.

TETRALOPHA MELANOGRAMMOS Zell..

The type of this species is a ♂ in the Cambridge Museum and through the kindness of Mr. Henshaw we have received an excellent photograph of the same together with the information that the secondaries are "pale whitish yellow, somewhat darker towards outer margin". We have identified a series from Kerrville, Texas, as belonging to this species; our ♂'s usually have a fairly distinct s. t. line on secondaries; only the faintest trace of this shows in Zeller's type, but otherwise our series matches excellently.

TETRALOPHA SUBCANALIS Wlk.

The single type in the British Museum is without head or abdomen and is worn, but the secondaries appear evenly smoky; we think it is a ♀ ; it agrees in maculation with a ♂ specimen in the Berlin Museum from Dallas, Texas, marked by Zeller *melanogrammos* but this specimen has the secondaries *evenly smoky* and should, we think, be referred to *subcanalis* Wlk.

In the Hulst collection the single ♀ type of *taleolalis* labelled "Col." is also listed as a synonym of *melanogrammos* and agrees with the specimen that we had compared with Zeller's specimen in Berlin. We have several ♂'s from South Pines, N. C., which we consider to be *subcanalis* Wlk.; these all show the dark secondaries. The maculation of the primaries in *subcanalis* and *melanogrammos* is extremely similar and the two names may apply to but one species; in view of the fact that the secondaries of the ♂'s show a color difference and that little is known of the life-history, we prefer to keep the names separate for the present. *Querciella* B. & McD. is considerably smaller but in markings is also very close to *subcanalis* Wlk.; in this instance we would concur with Dr. Dyar in his synonymy; we should however like to point out that the "sorry pile of synonymy" he accuses us of creating has dwindled down to three *sure* synonyms and one still doubtful one (*Homoeosoma differtella*); one of these (*Aresia parva*) would have taken priority over *ydatodes* Dyar if we had not delayed publication for over three months in a vain endeavor to induce Mr. Kearfott to publish descriptions of several species of Pyralids from Mr. Grossbeck's material so that we might use the names. Dr. Dyar's criticism therefore (Ins. Ins. Menst. I, 102) recoils on his own head and it is evident that in his haste to point out *our* gross ignorance "many subjects have not been carefully considered" by himself.

Dr. Dyar's synonymy of *melanogrammos* Zell. (Proc. Ent. Soc. Wash. VII, 30) shows the same hasty judgment as he displayed in his treatment of *militella* Zell; he lists *euphemella* Hulst, *tiltella* Hlst. and *speciosella* Hulst as synonyms, none of which species are referable to this name at all; *euphemella,* the type of which we have seen in the Hulst Coll. is the same species as *variella* Rag. and has a month or two priority; concerning this latter species we examined carefully the types of both this and *melanographella* Rag. at Paris and could find nothing definite whereby to separate them; we consider the synonymy as it stands in our lists to be correct.

Tiltella Hulst is a good species and close to *fuscolotella* Rag. being rather browner; we have a long and very constant series from Brownsville, Texas, which we compared with Hulst's type from Blanco Co., Texas, and found to agree.

Regarding *speciosella* Hulst we are in doubt, not having seen the type which we believe is in the National Museum.

TETRALOPHA ROBUSTELLA Zell. (syn. *diluculella* Grt.)

Hampson's synonymy as quoted by Dyar is correct; Zeller's types are in Berlin and Grote's in the British Museum.

TETRALOPHA TEXANELLA Rag.

This is not a synonym of *variella* Rag. as Dyar would have us
believe (Proc. Ent. Soc. Wash. VII, 33) nor of *subcanalis* as at
present listed; it is a good species rather resembling a large *callipeplella* Hulst with no fine dentations in the post median line and a general brownish tinge to base of primaries; the secondaries are pale.

This whole group will need further careful revision, especially
structurally and generically, but we would offer the following list of
species of the genus *Tetralopha* in its broad sense with tentative
synonyms in the hope of inducing a more careful study of these
interesting species:

(1) militella Zell.
    (? *platanella* Clem.)
(2) robustella Zell.
    diluculella Grt.
(3) asperatella Clem.
    expandens Wlk.
    (a) ? nephelotella Hlst.
      (*clemensalis* Dyar)
(4) aplastella Hulst.
(5) melanogrammos Zell.
(6) subcanalis Wlk.
    taleolalis Hlst.
    querciella B. & McD.
(7) floridella Hlst.
(8) speciosella Hlst.
(9) fuscolotella Rag.
(10) tiltella Hlst.
(11) humerella Rag.
    formosella Hlst.
(12) tertiella Dyar.
(13) baptisiella Fern.
(14) euphemella Hlst.
    variella Rag.
    melanographella Rag.
(15) slossoni Hlst.
(16) texanella Rag.
(17) callipeplella Hulst.

PHYCITINAE

Acrobasis Zeller.

In the Proc. Ent. Soc. Wash. X, 41 Dr. Dyar has revised this genus, giving a key to the species based on the ♂ sexual characters. Unfortunately for the general entomological public, who are apt to regard work from Washington as authoritative, Dr. Dyar has totally misidentified most of the older species, the types of which were not in America; it is a pity that an entomologist of Dr. Dyar's standing, with the ability and the opportunity to do such excellent work, has not ere this taken the trouble to obtain correctly determined material, but has relied on old existing determinations in the museum. In consequence a large portion of his work, as in the paper under discussion, is full of flaws, and only serves to make the existing confusion in nomenclature worse confounded; it is such work which is responsible for the constant shifting of names and the corresponding disgust on the part of economic workers with the whole question of nomenclature We offer the following notes based on the study of the type materia and would recommend to Dr. Dyar a similar method of procedure before he treats us to any further Keys to Species.

ANGUSELLA Grt. (syn. *eliella* Dyar)

In his original description of this species (N. Am. Ent. I, 51) Grote had two species confused; these he separates later, (Pap. I, 14) limiting *angusella* to the species with single black costal streak on underside of secondaries, a limitation confirmed by the ♂ type in the British Museum. According to Dr. Dyar's table *angusella* has *two* streaks on secondaries, which is wrong. A correct use of the table would lead us to *eliella* Dyar which is indeed synonymous with the true *angusella* Grt.; we have several specimens of *eliella* ex Coll. Merrick received from Mr. Ely himself and collected in the type locality.

DEMOTELLA Grt.

The type ♂ has a distinct black costal streak on under side of primaries, longer than in *angusella* and two streaks on secondaries, the costal one narrow at base and swelling to an oval patch at middle of wing, the median streak rather diffuse; this is probably Dyar's *angu-*

sella according to the key; we have 2 ♂'s in Coll. Barnes from Decatur, Ill., and New Brighton, Pa., which agree with type; the species is readily recognizable by the pale reddish basal area, including tuft, followed by a dark median band which shades into reddish terminally.

CARYAE Grote.

According to Dyar's table this species is unmarked in the ♂ sex beneath; the actual type lot of specimens in the British Museum shows that the ♂ has a short costal black streak and a longer median diffuse streak through the cell on primaries with secondaries unmarked; the true *caryae* is evidently what Dyar has identified as *caryivorella* Rag. a much larger species with no trace of red on primaries; the type of this latter species being a ♀, we had no means of determining the sexmarks of the ♂. *Caryae* Grt. was described from material from Mr. Coquillett and Illinois is the type locality (Bull. Geol. Surv. VI, 590). This spring we bred quite a series from larvae boring in the base of the young leaf-stems of shell-bark hickory around Decatur; the species is rather dark black-brown with a narrow oblique red line bordering the dark scale patch in lower half of wing; the sex dash in the ♂ on underside is sometimes rather indistinct and suffused when the ground color of the wing is deeper than usual, but is mostly quite distinct.

NEBULELLA Riley.

We would call Dr. Dyar's attention to the fact that Riley in the original description states (4th. Mo. Rep. p. 42) that the *single* type was bred from *crab-apple;* this hardly coincides with Dyar's statement (Proc. Ent. Soc. Wash. X, 45) that among the specimens of the supposed *nebulella* before him "4 bred by Dr. Riley on *hickory and walnut* including the type of *nebulella*" were included; we wonder if some of Riley's specimens of *juglandis* LeBaron could have led to this error; in any case it behooves Dr. Dyar to investigate the matter and give us some light on the true type of *nebulella* Riley.

PYLA AENEOVIRIDELLA Rag.

The type in the Ragonot Coll. bears the label "Evanston, Wyo., 14/6, 85", *not* N. Y. as erroneously given by Ragonot. One can easily understand how "Wyo." could have been mistaken for "N. Yo." by one not familiar with American states.

HOMOEOSOMA MUCIDELLUM Rag.

No type specimen is marked in Coll. Ragonot; the specimen figured however (Pl. XXXII, fig. 15) has been so labelled and is from Calif. (Wlshm.); as this is the locality given in the original description we may regard Ragonot's figure as that of the type.

INDEX

	Page
abbreviata Wlk.	209
abrasaria H. S.	204
abruptata Wlk.	207
aeneoviridella Rag.	222
aeria Grt.	201
affirmata Gn.	204
afflicta Grt.	200
albertina Hamp.	200
albigenaria Wlk.	211
albistrigalis B. & McD.	216
alciphearia Wlk.	212
altera Sm.	201
amatalis Wlk.	214
amoenaria Gn.	212
amyrisaria Wlk.	212
angusella Grt.	221
angustiorata Wlk.	209
anoa Dyar	201
anticaria Wlk. (Aethaloptera)	211
anticaria Wlk. (Eucymatoge)	203
antonita Dyar	201
aplastella Hlst.	220
arcasaria Wlk.	213
areaxaria Broad.	211
asperatella Clem.	218, 220
astarte Dbldy.	202
astylusaria Wlk.	212
atomaria Gn.	206
atrosignata Wlk.	209
auranticolor Grt.	198
baptisiella Fern.	220
barnesi Sm.	199
basaliata Wlk.	204
bistriolata Zell.	204
bitactata Wlk.	209
borealis Pack.	215
callipeplella Hlst.	220
calpota Sm.	198
canosaria Wlk.	210
capitalis Grt.	214
carcharodonta Hamp.	201
caryae Grt.	222
caryivorella Rag.	222
cervinifascia Wlk.	205

	Page
chamoechrysaria Grt.	206
chandleri Grt.	198
chapmani Grt.	202
claudialis Wlk.	215
clemensalis Dyar.	218
clitosalis Wlk.	202
colorado Sm.	198
comptaria Wlk.	203
concinna Warr.	214
condensata Wlk.	203
confusaria Hbn.	209, 212
confusalis Wlk.	215
congregata Wlk.	204
conocharodes Hamp.	201
consortalis Dyar	216
constricta Warr.	213
contemptata Gn.	208
continuata Wlk.	208
criminosa Swett	209
cruralis Gn.	202
cunigerata Wlk.	205
decisaria Wlk.	212
deductaria Wlk.	212
defectaria Gn.	211
definita Pack.	203
defluata Wlk.	209
demotella Grt.	221
densa Wlk.	200
densellus Zell.	217
deplanaria Wlk.	212
destinata Moesch.	205
diluculella Grt.	219, 220
discospilata Wlk.	206
dispuncta Wlk.	208
duaria Gn.	212
dubitata L.	204
duplicata Pack.	208
effectaria Wlk.	212
ejectaria Wlk.	211
eliella Dyar	221
ephyraria Wlk.	211
eremiata Gn.	207
eriosata Wlk.	210
euphemella Hlst.	219, 220

	Page
exauspicata Wlk.	207
excisa Wlk.	213
exnotata Wlk.	208
expandens Wlk.	218, 220
explagiata Wlk.	204
explanata Wlk. (Eucymatoge)	203
explanata Wlk. (Eustroma)	205
expressaria Wlk.	211
factiosalis Wlk.	202
fannia Druce.	201
faxoni Minot	206
fernaldi Grt.	209
fervidaria Hbn.	211
fidoniata Wlk.	206
fissinotata Wlk.	208
fistula Harv.	198
floridella Hlst.	220
fluctuata L.	205
formosella Hlst.	220
frigidata Wlk.	204
fulicalis Clem.	215
fumeola Hamp.	199
fumosaria Stkr.	210
furcifera Gn.	200
fuscolotella Rag.	218, 220
gargantua Dyar.	201
gortynoides Wlk.	200
grandis Hlst.	204
granitata Gn.	207
grotei B. & McD.	199
grotei Butl.	200
haliata Gn.	208
haliata Wlk.	208
hamamelis Gn.	200
harveyata Tayl.	205
heliothidata Gn.	208
hirtalis Gn.	214
humarium Gn.	210
humerella Rag.	220
ideata Wlk.	205
ignavaria Pears.	210
illaudata Wlk.	210
imitata Wlk.	211
implicata Wlk.	203
impulsa Gn.	200
incisa Harv.	213
incisaria Wlk.	212
inclinataria Wlk.	203
incrustalis Hlst.	217
indubitata Grt.	204
inextricata Wlk.	206
infumataria Grt.	210
inordinaria Wlk.	208
intermediata Gn.	205
intertexta Wlk.	211
intractaria Wlk.	210
intraria Gn.	211
invexata Wlk.	211
irregulata Wlk.	208
itysalis Wlk.	214
jacchusalis Wlk.	202
juglandis LeBar.	222
julia Hlst.	207
kentaria Grt.	213
lacustrata Gn.	205
lateritia Hbn.	200
latiorata Wlk.	209
leucostigma A. & S.	203
lobeliae Gn.	200
lucata Gn.	203
luteolata Hlst.	205
mactatoides B. & McD.	199
madusaria Wlk.	212
marcidilinea Grt.	202
matronalis Grt.	215
medicinalis Grt.	215
melanographella Rag.	219, 220
melanogrammos Zell.	218, 220
meskei Sm.	201
militella Zell.	217, 220
mira Dyar.	213
mirandata Tayl.	205
mixta Grt.	200
modiola Grt.	199
molliculata Wlk.	205
momaria Wlk.	211
monodia Dyar.	202
mucidellum Rag.	223
muiri Hy. Edw.	214
muzaria Wlk.	212
nasonialis Zell.	214
nebulella Riley	222
nephelotella Hlst.	218, 220
nexalis Hlst.	214
nigrocomina Warr.	208
nivea Wlk.	216
notataria Wlk.	206
nuicola Sm.	201
nymphulalis Haim.	214
obliqua Pack.	203

Page

obliquata Hy. Edw.......... 216
obliviosa Wlk. 200
obtusaria Hbn. 212
ocellinata Gn. 208
oecogenes Dyar. 200
ophionalis Wlk. 214
oponearia Wlk. 212
ordinata Wlk. 207
orillata Wlk. 208
ornata B. & McD.......... 213
orthozona Hamp. 201
ovulalis Gn. 213
paenulata Clem. 213
pampinaria Gn. 210, 211
pectinaria D. & S. 212
pellucidaria G. & R. 211
pellucidaria Pack. 210
perangulata Hlst. 212
perlubens Grt. 198
perseella B. & McD.......... 217
picta Feld. 203
piniata Pack. 209
plagiata Wlk. 203
platanella Clem. 217, 220
platinalis Gn. 213
populata L. 205
probata B. & McD. 199
progressata Wlk. 204
propriaria Wlk. 212
propulsata Wlk. 205
proxanthata Wlk. 208
pterophoralis Wlk. 215
pustularia Hy. Edw......... 204
pyrolaria Gn. 210
quadrisignata Wlk. (Homopyralis) 202
quadrisignata Wlk. (Macaria) 208
quadristrigalis Fern. 214
querciella B. & McD. 219, 220
remotata Wlk. 205
renunciata Wlk. 204
repugnatalis Wlk. 216
restorata Wlk. 208
retinotata Wlk. 208
retractaria Wlk. 208
rilevaria Pack. 206
rivulalis Hamp. 214
robiginosus Morr. 210
robustella Zell. 219, 220
rubralis Warr. 214
rubrica Harv. 198

Page

santa rita Dyar. 201
schistacea Warr. 205
scirpicola Gn. 198
scitata Wlk. 211
scolopacinaria Gn. 210
seductaria Stkr. 209
semiclusaria Wlk. 210
serrata B. & McD. 202
similis Wlk. 205
simplex Warr. 215
sirenaria Stkr. 212
slossoni Hlst. 220
snoviaria Pack. 206
solanalis B. & McD. 215
speciosella Hlst. 219, 220
strigularia Wlk. 200
subapicalis Sm. 198
subapicaria Wlk. 208
subcanalis Wlk. 218, 220
subcolumbata Wlk. 208
subgenericata Dyar. 211
sublunaria Gn. 211
submuraria Wlk. 211
subsequalis Gn. 215
succandidalis Hlst. 215
succosata Zell. 208
sulphurata Pack. 213
taeniata Steph. 204
takenaria Pears. 211
taleolalis Hlst. 218, 220
tantillus Grt. 202
tertiella Dyar. 220
texanella Rag. 220
tillialis Dyar. 214
tilosalis Sm. 202
tiltella Hlst. 218, 219
tiviaria Wlk. 212
transfigurata Swett. 204
transfixaria Wlk. 211
turmalis Grt. 214
ulamora Sm. 198
unangulata How. 204, 205
unicolor B. & McD. 216
unimodaria Morr. 207
varadaria Wlk. 209
varia Wlk. 203
variegata Wlk. 214
variella Rag. 219, 220
xanthindyma Bdv. 202
xanthometata Wlk. 206
zalissaria Wlk. 209

CONTRIBUTIONS

TO THE

NATURAL HISTORY

OF THE

LEPIDOPTERA

OF

NORTH AMERICA

VOL. II
No. 6

SOME NEW NORTH AMERICAN PYRAUSTINAE

BY

WILLIAM BARNES, S. B., M. D.

AND

J. H. McDUNNOUGH, Ph. D.

DECATUR, ILL.
THE REVIEW PRESS
AUGUST 12, 1914

Published
Under the Patronage
of
Miss Jessie D. Gillett
Elkhart, Ill.

PREFACE

For several years a number of unidentified species of the subfamily Pyraustinae have been accumulating in the Barnes Collection. Recently on a visit to the British Museum we took specimens of these species with us in the hopes of getting them identified; in some few instances we were successful and several of these names proved to be new to our faunal region; in the majority of cases however we failed to find any names applicable and venture in the following paper to describe these species as new. We are fairly certain at all events that they have never been described previously from N. American material, as only some half dozen of the species at present listed in Dyar's catalogue are unknown to us and the descriptions of these do not appear to fit any of our material. Of course there is always the possibility that some of our species will prove identical with Mexican or West Indian species, but we have no doubt that those better placed than we are to study such material will find no difficulty in correcting any errors into which we have fallen on this score.

We preface our descriptions with a list of species which should be added to our present list.

SPECIES NEW TO OUR LIST

EURRHYPARODES LYGDAMIS Druce.

A series of both sexes from Brownsville, Tex., agrees with a figure of Druce's type in the British Museum. In the ♂ the secondaries are considerably yellower than in the ♀.

GONOCAUSTA ZEPHYRALIS Led.

We have a single specimen of this species from Brownsville, Texas; we cannot find either genus or species in Hampson's revision, our copy being without an index, so do not know if Lederer's name has priority or not.

CINDAPHIA ANGUSTALIS Feld.

A long series of this species from Brownsville and San Benito, Texas, is contained in Coll. Barnes.

ANTIGASTRA CATALAUNALIS Dup.

We have a series of this widespread species from Redington, Ariz.

PYRAUSTA TAENIOLALIS Gn.

We have a series of this species from Brownsville and San Benito, Texas; the identification is according to the British Museum material and appears to fit very well with Guenée's description. We see no reason for retaining this in *Pionea* as Hampson has placed it; it seems much better placed between *phoenicealis* Hbn. and *onythesalis* Wlk. (*insignatalis* Gn. a/c Brit. Mus.)

DESCRIPTIONS OF NEW SPECIES

GLAPHYRIA DUALIS sp. nov. (Pl. I, Fig. 1).

Head, thorax and antennæ pale straw-color; primaries pale ochreous variably sprinkled with brown scales which at times almost obscure the ground-color, crossed by two prominent dark lines; t. a. line blackish, outwardly oblique to below cubitus, then angled and inwardly oblique to inner margin; t. p. line blackish, bent inwards at costa, rounded outwardly opposite cell, parallel to t. a. line in lower portion, faintly bordered with white outwardly; reniform a dark dash; terminal row of black dots, fringes smoky, whitish outwardly. Secondaries whitish, tinged with ochreous terminally, with faint dark discal dot and curved subterminal line; terminal row of black dots; fringes deep smoky, whitish-tipped. Beneath whitish, tinged with smoky with maculation of upper side repeated, t. p. line being especially prominent. Expanse 16 mm.

HABITAT. San Benito, Texas (Mch. 16-23); Brownsville, Texas; 9♂, 3♀. Types, Coll. Barnes.

Closely allied to *dichordalis* Hamp. of which species we examined the type in the British Museum. The primaries are slightly subfalcate with rather prominent bulge in central portion of outer margin.

ERCTA DESMIALIS sp. nov. (Pl. I, Fig. 2).

Palpi brown above, white beneath; head and thorax brown; primaries brown over a whitish ground-color which is only apparent under the lense; a quadrate white patch near end of cell, resting on cubitus and bordered laterally by black lines; below this a second similar patch, the upper margin of which touches cubitus at inception of vein 2, the lower margin resting on vein 1; a 3rd white irregularly oval patch occupies the sub-terminal area between veins 3 and 8, is outlined in black and connected with costa by a dark streak; fringes light smoky, dotted with white at base and cut by a dark line, above anal angle they are white. Secondaries brown crossed in the median area by two semi-parallel waved dark lines, the outer of which does not attain inner margin; the space between these lines is whitish sprinkled with brown scales; fringes as on primaries. Beneath essentially as above, but secondaries paler, being in some cases almost entirely white. Expanse 17-20 mm.

HABITAT. Palmerlee, Ariz. 6 ♂'s, 6 ♀'s. Types, Coll. Barnes.

Besides the above types there is a long series before us from the same locality; the 3rd joint of the palpi we would consider to be short and blunt, not triangularly tufted as given in Hampson's Key as one of the characteristics of the genus *Ercta;* in other respects however, viz., the annulate antennae, the minute maxillary palpi, the triangularly scaled 2nd joint of the labial palpi and the general frail appearance, the species fits in so well with the definition that we place it in *Ercta;* the maculation is rather Desmiid in character.

HEDYLEPTA? FUTILALIS sp. nov. (Pl. I, Fig. 3).

Palpi, head, thorax and abdomen clear yellow, rear segments of latter posteriorly narrowly bordered with white, anal segment in ♂ with centro-dorsal black streak, bordered laterally with white, beyond which is smoky shading, anal hair pencil of a pale ochreous color; in ♀ small black anal tuft bordered anteriorly with white. Primaries clear yellow, slightly deeper outwardly; t. a. line black, fine, gently rounded outwardly; no orbicular; reniform a black dash; t. p. line black, from a point on costa half way between reniform and apex of wing, almost straight to vein 3, then inward and almost obsolete to just below reniform and again straight and distinct to inner margin; black terminal line; fringes dusky, cut by a darker line near base.   Secondaries similar in color to primaries with blackish discal dash continued to inner margin just above anal angle by a black line; beyond the cell a slightly curved black line from costa to vein 2; outer margin of fringes as on primaries.   Beneath paler than above with similar markings and slight smoky shade terminally.   Expanse 14 mm.

HABITAT. San Benito, Texas (Aug.); Brownsville, Tex. 1 ♂, 4 ♀. Types, Coll. Barnes.

The palpi have the 3rd joint porrect, scarcely distinguishable, lying on the tufted 2nd joint; the species is very close in maculation to *indicata* Fabr., the type of the genus *Hedylepta,* which shows however a tufted patagium in the ♂, a feature lacking in our species; the abdominal markings are also different. According to Hampson's Key this species would fall under *Cliniodes,* but we do not like the association and prefer to place it provisionally in *Hedylepta* Led.

BLEPHAROMASTIX SANTATALIS sp. nov. (Pl. I, Fig. 5).

Palpi brownish outwardly, mixed with white on 3rd joint; front ochreous, lined laterally with white; vertex of head ochreous, collar white; thorax white with broad ochreous band anteriorly, abdomen ochreous, ringed with white; primaries white suffused with ochreous; an inwardly oblique dark basal line suffused along its edges with pale ochreous; t. a. line brown, slightly outwardly oblique to cubitus, then slightly bent outwardly and perpendicular to inner margin; t. p. line brown, almost parallel to outer margin from costa to below vein 3, then rounded and bent strongly inwards and upwards to origin of vein 3, again rounded and inwardly oblique to inner margin, very close to or even touching t. a. line at this point; orbicular a large oval, filled with pale ochreous and partially outlined with brown; reniform a large quadrate ochreous patch, slightly oblique, outlined laterally with brown, the inner line touching bend of t. p. line;

median space largely ochreous shaded, leaving white area between the spots and before and behind same; a dark s. t. line close to outer margin and sending short spurs along the veins towards same, preceded by ochreous shading; terminal area white with dark terminal line; fringes ochreous in basal half, white outwardly, the two areas separated by a dark line. Secondaries white, ochreous shaded, with dark t. p. line of primaries practically duplicated, straight from costa to a point near outer margin below vein 2, then bent strongly backwards to origin of vein 3, continued along cubitus to origin of vein 2, then to inner margin above and angle, shaded inwardly with ochreous, which fills the outer projection; marginal maculation and fringes as on primaries. Beneath essentially as above, but paler. Expanse 15 mm.

HABITAT. San Benito, Texas (Apr. 24-30), 3 ♀. Type, Coll. Barnes.

Falls into the same group with *aplicalis* Gn. which also has the palpi with 3rd joint porrect and resting on the triangular tufting of joint 2. This feature is found in the type of the genus, *ranalis* Gn. which Hampson incorrectly places in *Nacoleia* with 3rd joint upright and triangularly tufted.

SYLEPTA BRUMALIS sp. nov. (Pl. I, Fig. 4).

Antennae in ♂ normal; fore femora, tibiae and tarsi in ♂ strongly tufted with hair, in ♀ femora and tibiae slightly tufted; mid tibiae slightly tufted in ♂; hair tuft between the antennae stronger in ♂ than ♀; palpi, head and thorax light brown mixed with white; 1st joint of abdomen white, remainder light brown; hair on basal joints of fore legs brown, tarsi white; primaries with costal margin pale yellowish narrowing towards apex; base of wing light brown containing a white dot near inner margin and bordered outwardly by pale yellowish perpendicular band, scarcely 2 mm. broad; remainder of wing dirty olive-green with the exception of a portion of the cell which is white containing a broad lunate olive green reniform; a faint darker t. p. line outwardly rounded opposite cell, incurved in fold, shaded slightly outwardly at costa with pale yellowish. Secondaries white in basal third with small olive green costal dot; following this is a broad olive green band the outer margin of which forms the continuation of the t. p. line of primaries and is shaded outwardly, especially above anal angle, with white; remainder of wing olive-green as on primaries, slightly paler than the transverse band; a slight brownish shade at anal angle; fringes concolorous, paler outwardly and whitish at anal angle of secondaries. Beneath hyaline whitish with markings of upper side more or less distinguishable. Expanse ♂ 25 mm., ♀ 32 mm.

HABITAT. Brownsville, Texas; San Benito, Texas (Dorner), March-May and July-Sept. 9 ♂'s, 9 ♀'s. Types, Coll. Barnes.

Seems best placed in *Sylepta*, although the palpi in tufting are about intermediate between this genus and *Pilocrocis*, being slightly less upturned than in our other N. Am. species of either genus. The hair· tufts on the legs of the ♂ combined with the simple antennae would throw it into Section VII B of Hampson's revision (Proc. Zool. Soc. Lond. 1898, p. 716).

HELLULA AQUALIS sp. nov. (Pl. I, Fig. 6).

Head and thorax pale ochreous; primaries pale ochreous shaded with olivaceous; basal space to t. a. line evenly olivaceous; t. a. line white, oblique to below cubitus, then straight to inner margin, bordered outwardly by an olive-brown hair line; reniform oblique, constricted in middle, filled with smoky-brown, median space largely pale ochreous, except around reniform where it is slightly olive shaded; t. p. line white, finely dentate, well rounded outwardly opposite cell, incurved in fold with small outward tooth on vein 1, the whole line bordered inwardly by an olive-brown line, slightly broader at costa; terminal space olivaceous; obsolescent row of terminal black dots; fringes olivaceous in basal half, whitish outwardly. Secondaries semihyaline white with faint smoky terminal border; fringes white. Beneath paler than· above with the markings repeated. Expanse 16 mm.

HABITAT. Santa Catalina Mts., Ariz. (Sept.) (♂); Redington, Ariz. (♀). Types, Coll. Barnes.

The ♀ is much paler than the ♂ with the maculation obsolescent. The 3rd joint of the palpi is shortly triangularly scaled and the species would thus not fall into *Hellula* by Hampson's tables. An examination of our N. Am. species going as *undalis* Fabr. shows however that our new species is very closely allied structurally to it and in maculation has also considerable general resemblance.

Genus EVERGESTIS Hbn.

There are a number of closely allied species belonging to the *funalis-obliqualis* group which have apparently been overlooked by systematists. We have had them separated for some time, but have not ventured to describe them as we were uncertain as to our identifications of the species already named. Our visit to the British Museum allowed us to exactly match *funalis* Grt.; the type is a Henry Edwards' specimen from the Lake Tahoe region of the Sierra Nevadas, Calif., and we possess a very perfect pair from the same locality. (Pl. II, Fig. 1.) From the Coll. Merrick we obtained three specimens of the type lot of *obliqualis* Grt. received originally from Prof. Snow, which confirmed our identification of this species (Pl. II, Fig. 2); the type of *simulatilis* Grt. was also seen in the British Museum and matched from our material. *Napaealis* Hlst. should be removed from *Evergestis* and would probably fall in *Pyrausta;* we saw the type in the Hulst Collection and have a long series from San Diego, Calif. The following species appear to be undescribed.

EVERGESTIS INSULALIS sp. nov. (Pl. II, Fig. 4).

Head, thorax and abdomen gray; primaries whitish-gray sprinkled at base and along costa with purplish-gray and shaded terminally with dull red-brown and smoky; t. a. line very faint, often obsolete,. twice dentate below costa, then inwardly oblique and almost straight to inner margin, ending in a small distinct dark spot; reniform an obscure quadrate dark blotch, its lower end only slightly removed from t. p. line; this line rather faint at costa, then broad, dark brown, and very distinct, twice dentate below costa then inwardly oblique and parallel to t. a. line with a slight incurve below reniform; subterminal line smoky, diffuse, close to margin with several prominent dark dashes below apex of wing extending to t. p. line and a smoky patch above inner margin, touching t. p. line and leaving a short vertical white dash exterior to t. p. line resting on inner margin; between the apical dashes and the anal patch the subterminal space is shaded with dull red-brown; faint dotted dark terminal line; fringes with ochreous basal line followed by a deep smoky line, outer portion pale smoky mixed with white. Secondaries semihyaline whitish, shaded outwardly with smoky with a dotted smoky subterminal line; fringes as on primaries. Beneath whitish with slight smoky tinge and markings of upper side showing through. Expanse 24-28 mm.

HABITAT. Duncans, Vanc. Is., B. C. (Aug.-Sept.); Quamichan Lake, Vanc. Is.; Victoria, Vanc. Is. 2♂, 5♀. Types, Coll. Barnes.

Closely related to *funalis* Grt., but median area paler, maculation less sharply defined and dark markings above anal angle more diffuse, forming a patch, *not dashes,* which reaches to t. p. line; secondaries paler.

EVERGESTIS TRIANGULALIS sp. nov. (Pl. II, Fig. 5).

Head and thorax light purplish gray; primaries whitish heavily scaled with brown in basal and median areas, leaving only traces of ground color visible, and suffused with dull red-brown terminally; t. a. line faint, red-brown, inwardly oblique, non-dentate, preceded by a white shade usually more distinct than line itself; t. p. line toothed below costa, then evenly oblique to inner margin and parallel to t. a. line; prominent white shading before t. p. line; reniform an obscure elongate cloud; slight dark shade on inner margin at inception of t. a. line; beyond t. p. line and resting on inner margin a prominent deep red-brown triangular patch bordered outwardly by a whitish line followed by gray shading extending to outer margin; terminal area above this rather even dull red-brown with faint traces of a smoky s. t. line; faint terminal dotted line; fringes smoky with paler basal, median and terminal lines. Secondaries pale smoky hyaline, shaded with deep smoky outwardly, with dotted subterminal line and pale ochreous fringes shaded with smoky between veins 2 and 3. Beneath pale ochreous, overlaid with smoky, with dark subterminal line to both wings and prominent reniform on primaries. Expanse 27-30 mm.

HABITAT. Santa Catalina Mts., Ariz. (Sept.); Palmerlee, Ariz.; Redington, Ariz.; Chiricahua Mts., Ariz.; Yavapai Co., Ariz. 1 ♂, 8 ♀. Types, Coll. Barnes.

Allied to *obliqualis* Grt., from which it is readily distinguished by the sharply defined triangular patch above inner margin and the even t. a. line without dentations. Both this species and *obliqualis* are in Coll. Barnes, labelled *funalis* Grt. in Dr. Dyar's handwriting from material sent in several years ago to the National Museum for identification. We mention this to show the class of identification work to be expected from this source; this is by no means an isolated case, but only one out of many cases of misidentification, as shown by material in Barnes Collection.

EVERGESTIS SUBTERMINALIS sp. nov. (Pl. II, Fig. 7).

Basal area of primaries to t. p. line heavily overlaid with a mixture of blackish and white scaling, paling outwardly; t. a. line oblique inwardly, waved, twice toothed below costa, projecting outwardly on the fold, inwardly on vein 1, ending in dark spot on inner margin, preceded especially above inner margin by a white shade; reniform an obscure quadrate patch; t. p. line dark, dentate below costa, then inwardly oblique and parallel to t. a. line, ending in a dark streak on inner margin, preceded by light ochreous shading and followed above inner margin by a small ochreous lunate mark; subterminal area almost entirely bright red-brown except below costa where it is shaded with smoky; s. t. line diffuse, smoky, close to outer margin, defined outwardly by gray terminal shading and connected with t. p. line by fine dark lines above and below vein 6 and on vein 1; broad dark terminal line widening towards apex; fringes smoky, narrowly checkered with ochreous and cut by a median paler shade. Secondaries pale hyaline ochreous, shaded with smoky along outer margin with very faint subterminal dark line; fringes pale ochreous, cut by two darker lines. Beneath pale ochreous, primaries largely suffused with smoky, with common dotted s. t. line and large reniform on primaries. Expanse 31 mm.

HABITAT. Deer Park Spgs., Lake Tahoe, Calif.; Truckee, Calif. (Sept. 24-30). 2 ♀, Type, Coll. Barnes.

Very similar to *funalis* Grt. but larger, the basal areas darker, the subterminal area with more extended red-brown shade and lacking the smoky area above anal angle. It is possible Grote had both species before him when describing *funalis* as his figure (N. Am. Ent. I, Pl. V.) looks more like this species than the true *funalis* according to the specimen labelled "type" in the British Museum. There are several worn specimens from Glenwood Spgs., Colo., in Coll. Barnes that may be this species.

EVERGESTIS EUREKALIS sp. nov. (Pl. II, Fig. 3).

Primaries rather even lilac-gray; t. a. line often obsolete, when present inwardly oblique with prominent outward angle in cell and a smaller one on fold; reniform patch scarcely visible; t. p. line dark, especially above inner margin, dentate below costa, then evenly oblique and parallel to outer margin, preceded by light shading, followed by a narrow white line; subterminal area shaded with pale red-brown, especially in central portion; some dark apical dashes and dark shading above anal angle, connected by a faint diffuse dark shade representing the s. t. line; terminal space light gray; a dark terminal line; pale ochreous fringes cut by two smoky lines. Secondaries pale hyaline smoky with dark terminal border and subterminal line dotted on the veins; a broken dark terminal line; fringes with pale basal line followed by dark line, outwardly pale with slight smoky shade between veins 2 and 3. Beneath pale smoky with dark reniform mark on primaries and common dark s. t. line. Expanse 29 mm.

HABITAT. Eureka, Utah (Aug.-Sept.), (Spalding). 4 ♂, 11 ♀. Types, Coll. Barnes.

Very close in color and markings to *funalis* Grt. differing principally in the course of the t. a. line which in *funalis* shows none of the prominent angles found in our species; the t. p. line is more oblique and less bent in fold than in *funalis*.

EVERGESTIS VINCTALIS sp. nov. (Pl. II, Fig. 8).

Primaries pale olive brown shaded with bluish-gray; t. a. line brown, edged inwardly broadly with pure white, indistinct at costa, strongly inwardly oblique from a point near reniform, forming an inward tooth below vein 1; reniform a small obscure dark shade; t. p. line brown, broadly white-shaded inwardly, dentate below costa, then oblique and parallel to t. a. line forming a slight inward tooth on vein 2 and bent sharply outwards below vein 1 forming an angle on this vein; outwardly the t. p. line is edged very narrowly with white except at costa, where the white area is rather diffuse; s. t. line represented by faint irregular white line in lower half of wing; terminal area shaded with bluish-gray with terminal dark line edged inwardly with white; fringes pale cut by two darker lines. Secondaries pale smoky, slightly darker terminally with indistinct subterminal line and whitish subterminal markings on vein 2, separated by smoky dots; fringes similar to those of primaries, but paler. Beneath whitish with faint traces of common subterminal line. Expanse 26 mm.

HABITAT. Jemez Spgs., N. M. (Apr. 8-15); Glenwood Spgs., Colo. (May, Aug.). 2 ♂, 2 ♀. Types, Coll. Barnes.

Allied to *simulatilis* Grt., but lacks the prominent angle in t. a. line below costa of this species and is paler and smaller.

EVERGESTIS OBSCURALIS sp. nov. (Pl. II, Fig. 6).

Primaries dull smoky brown slightly powdered with whitish with macula-tion very obscure; t. a. line scarcely visible, inwardly oblique, irregularly den-tate; t. p. line obscure, dentate below costa then oblique and parallel to t. a. line with slight incurve in fold, bordered outwardly by faint white line; reniform a large quadrate dark blotch, almost resting on t. p. line; s. t. line a diffuse white shade, obsolete towards apex of wing, parallel to t. p. line; terminal broken dark line; fringes pale, cut by two dark lines, ochreous at base. Secondaries smoky, darker along outer margin with an obscure subterminal line bordered outwardly by a pale line not attaining anal angle; fringes as on primaries but paler. Beneath pale ochreous, clouded with smoky, with common subterminal line, dentate on secondaries and large reniform patch on primaries. Expanse 32 mm.

HABITAT. Silverton, Colo. (10,000 ft.) (Aug. 1-7) 3 ♂, 1 ♀. Types, Coll. Barnes.

Easily recognized by its dull coloration and obscure markings; it is evidently a high altitude form. The ♀ is rather better marked than the ♂'s and the space between s. t. and t. p. lines is slightly browner than the ground color.

EVERGESTIS ARIDALIS sp. nov. (Pl. II, Fig. 9).

Head, thorax and primaries whitish ochreous, shaded with darker; pri-maries lemon yellow crossed by two indistinct darker parallel lines, the inner strongly outwardly oblique to cubitus at inception of vein 2, then sharply angled and inwardly oblique to inner margin; outer line bent outwards gently below costa then parallel to the inner line; reniform faintly visible as a dark streak outlined with paler; a very faint dark shade line from apex of wing to t. p. line; crosslines slightly punctate on the veins. Secondaries whitish hya-line, slightly yellow along terminal border. Beneath shiny whitish. Expanse 25 mm.

HABITAT. Esmeralda Co., Nevada. 1 ♂, 1 ♀. Types, Coll. Barnes.

Really closer to the European extimalis Scop. than consimilis Warr. which has wrongly been listed as a synonym by Hampson; our new species differs in the pale ochreous fringes, not plumbeous, and the lack of the dark terminal shading of primaries; the t. p. line is also rather more incurved and dentate at costa; the primaries are narrower with more pointed apex. Extimalis should be dropped from our lists.

EVERGESTIS? LUNULALIS sp. nov. (Pl. II, Fig. 11).

Primaries hyaline whitish, shaded with pale yellow; basal area suffused with yellowish; t. a. line faint, inwardly oblique, slightly waved, ending in a slight dark spot on inner margin; reniform an obscure yellowish patch; t. p.

line yellowish, outwardly oblique to vein 8, then parallel to outer margin as far as vein 5, then incurved to vein 1 and forming a prominent lunule which is heavily shaded with brown and is the most striking feature of the maculation; beyond this lunule diffuse yellowish shading; apex of wing yellowish continued as a faint s. t. line to inner margin; fringes yellowish cut by a slightly darker line. Secondaries hyaline white, slightly smoky outwardly with faint curved postmedian line. Beneath whitish, primaries slightly smoky with well defined t. p. line and reniform spot, secondaries with faint postmedian line as above. Expanse 21 mm.

HABITAT. Palmerlee, Ariz.; Wilgus, Ariz. 2 ♂, 2 ♀. Types, Coll. Barnes.

Possibly better placed in *Crocidophora* than *Evergestis* as the 3rd joint of the palpi appears to be concealed. We place it for the present in *Evergestis* as it appears to bear affinities to the *rimosalis* group of this genus.

LOXOSTEGE TYPHONALIS sp. nov. (Pl. I, Fig. 7).

Head, thorax and primaries ochreous brown, the latter shaded in the median area beyond the cell and in the lower half with dark brown variably overlaid with pale blue scaling; t. a. line obscure, smoky, strongly outwardly oblique to above fold, then slightly bent inwards to inner margin; t. p. line outwardly rounded opposite cell, then strongly incurved, forming a tooth pointing inwardly on vein 2 and another on vein 1 and more than twice as close to t. a. line on inner margin as on costa; orbicular small, brown, oblique; reniform small, obscure, lunate, often hidden by a brown shade extending beyond the cell to t. p. line and sprinkled with blue; a faint dark marginal line; fringes concolorous. Secondaries pale ochreous, darker marginally with traces centrally of a waved postmedian line. Beneath pale ochreous, with markings of primaries on upper side showing more or less through. Expanse 18-20 mm.

HABITAT. Redington, Ariz.; Santa Catalina Mts., Ariz. (July); Babaquivera Mts., Ariz.; Christmas, Gila Co., Ariz. 8 ♂, 8 ♀. Types, Coll. Barnes.

Belongs in the *allectalis-albicerealis* group and is very close in maculation to *allectalis* Grt. differing in its light-brown color with much reduced blue shading.

LOXOSTEGE ROSEITERMINALIS sp. nov. (Pl. I, Fig. 8).

♂ antennae strongly ciliate; front oblique, roof-shaped, ending in a conical tubercle; head, thorax and primaries bright yellow, latter with a narrow terminal border and the fringes purplish-pink; no markings. Secondaries including fringes whitish. Beneath unicolorous whitish. Expanse 15-20 mm.

HABITAT. San Benito, Texas; Brownsville, Texas; (March-Apr., June, Sept.) 6 ♂'s, 6 ♀'s. Types, Coll. Barnes.

DIASEMIA LEUCOSALIS sp. nov. (Pl. I, Fig. 9).

Primaries light yellow with slight sprinkling of brown scales and veins more or less marked in brown; basal portion of costa shaded with purple-brown, t. a. line fine, brown, outwardly rounded on vein 1; orbicular round, more or less filled with purple-brown; reniform a quadrate purplish blotch the lower portion of which is usually lost in some brown scaling and the upper portion almost attains costa; a small quadrate hyaline white spot separates the reniform and orbicular; t. p. line brown, straight from costa to vein 5, then rounded inwardly to base of vein 2 and inwardly oblique to inner margin near t. p. line with slight outward curve across vein 1; the remainder of the wing beyond the t. p. line is largely deep purplish, mixed with brown, leaving three small yellow triangular spots next t. p. line between veins 5 and 8 and a narrow yellow band on outer margin below apex, divided into spots by the brown veins and usually not extending much below vein 5; below the cell the purple suffusion is bordered inwardly by a darker line extending from base of vein 3 to inner margin parallel to t. p. line and leaving a narrow band of clear yellow between the two lines, divided into patches by the brown veins; apex of wing shaded with brown without purple tinge; terminal dark line; fringes brown in basal half, whitish outwardly. Secondaries pale whitish-yellow with broad brown outer border, narrowing towards anal angle; a minute discal dot and a median dark line from cubitus to inner margin of wing parallel to inner margin of terminal band; fringes as on primaries. Beneath whitish with the markings of the upper side reproduced in smoky. Expanse 15-17 mm.

HABITAT. Palmerlee, Ariz.; White Mts., Ariz. 2 ♂, 4 ♀. Types, Coll. Barnes.

Allied to *erubescens* Hamp. if Druce's figure (Biol. Cent. Am. Pl. 101, Fig. 17) be correct, and also to *elegantalis* Warr.

DIASEMIA ZEPHYRALIS sp. nov. (Pl. I, Fig. 10).

Palpi blackish, whitish beneath at base; head and thorax, sulphur-yellow; primaries sulphur yellow shading into orange-yellow terminally, costa to reniform purple; orbicular oval, reniform figure of 8, both filled with purple and attached to costal stripe; traces of a t. a. line and a median line in a few purplish dots above inner margin; t. p. line deep purplish, finely dentate, almost straight below costa, bent sharply at right angles along vein 3 to below reniform and then again at right angles to inner margin, the sinus thus formed filled with a purple suffusion which extends narrowly upwards along outer edge of t. p. line and broadens out below costa to a second patch; fringes yellow. Secondaries similar in color to primaries; a subbasal dark line only prominent above inner margin, a small discal dot and a subterminal line continuous with the t. p. line of primaries and squarely exserted between veins 3 and 6; a purplish suffusion beyond this line especially prominent from vein 3 to anal angle.

Beneath pale ochreous with the maculation of the upper side more or less repeated in smoky-brown. Expanse 23 mm.

HABITAT. Mineral King, Tulare Co., Calif. (July), (G. R. Pilate). 3 ♂, 5 ♀. Types, Coll. Barnes.

Allied to *plumbosignalis* Fern., but readily distinguished by its bright yellow secondaries.

DIASEMIA? FENESTRALIS sp. nov. (Pl. I, Fig. 11).

Antennae of ♂ annulate; labial palpi long, beak-like, 3rd joint slightly bent downward at extremity; maxillary palpi tufted with scales at extremity, front rounded, not prominent; primaries hollowed out below apex and again above anal angle, forming a rather prominent bulge between veins 2 and 5; secondaries slightly excised below apex; head, thorax and primaries light brown, latter with the median area more or less broken up into semihyaline white patches, defined by the dark veins and cross lines; costa and terminal area brown; wing crossed by three fine brown lines, the first subbasal only distinct in median area where it is preceded by a white blotch, the second antemedial, oblique outwardly to cubitus, then parallel to outer margin, the third postmedial, outwardly oblique to vein 7 then rather irregularly waved and semi-parallel to outer margin; orbicular occasionally present as a brown dot encircled with whitish; reniform a brown lunate mark; beyond the third line the intravenular spaces contain a row of whitish blotches separated by the brown veins and between veins 3 and 5, beyond this row, are two more similar blotches with an occasional third one between veins 2 and 3; fringes black brown, white in the excisions. Secondaries hyaline white with faint trace of a dotted postmedian line and a slightly broken terminal dark line; fringes smoky in costal half, whitish towards anal angle. Beneath whitish hyaline, primaries with an irregular dark smoky border, a dark reniform and a dark dash on costa at inception of 3rd cross line of upper side, secondaries with a rather prominent dark costal blotch ⅓ from apex and a smaller one near base; a dotted dark terminal line and traces of the subterminal line of upper-side. Expanse ♂ 23 mm.; ♀ 28 mm.

HABITAT. Palmerlee, Ariz.; Santa Catalina Mts., Ariz.; 2 ♂, 4 ♀. Types, Coll. Barnes.

According to Hampson's tables the species would fit equally well into half a dozen genera; we place it in *Diasemia* on the strength of the annulate ♂ antennae and the excised apex of secondaries; the shape of the primaries is much as in *oblectalis* Hulst, but the frontal structure would preclude association with this species.

GYROS (MONOCONA) ATRIPENNALIS sp. nov. (Pl. I, Fig. 12).

Palpi, head and thorax covered with long greenish white hair; fore and mid-legs white, hind legs black; primaries blackish with a considerable deep purplish-red tinge; basal half of wing defined outwardly (at times quite prominently) by a curved dark t. a. line just beyond which a black dash in the cell represents the reniform; t. p. line blackish, very close to outer margin with which it is practically parallel; the space between these two lines sprinkled with white scales, often quite heavily; fringes blackish, slightly ochreous at tips. Secondaries and underside deep-black; abdomen black, at times with posterior half tinged with reddish. Expanse 13 mm.

HABITAT. Mineral King, Tulare Co., Calif. (10,000 ft.) (Pilate). 6 ♂, 5 ♀. Types, Coll. Barnes.

This may be a variety of *muiri* Hy. Edw. (*rubralis* Warr.) of which we possess a long series from the type locality, Truckee, Calif.; two specimens of the new species show a distinct reddening of the secondaries both above and below without however approaching the bright color of *muiri*. The t. p. line in *atripennalis* appears to be closer to the outer margin and more evenly curved than in *muiri*, so for the present we prefer to regard it as a distinct species.

ORENAIA COLORADALIS sp. nov. (Pl. I, Fig. 13).

Palpi, head and thorax blackish, mixed with white scaling; primaries blackish heavily scaled with whitish; a black dot at base of wing followed by an obscure black sub-basal line with outward tooth on median vein; t. a. line black, bordered inwardly with white, almost perpendicular, with prominent dentations on cubitus and vein 1; reniform a small black dash; t. p. line black, bordered outwardly with white, finely dentate, slightly outcurved around cell; terminal area darker than median; fringes smoky. Secondaries smoky with an obscure dark subterminal line bent inwards on vein 1 and a faint discal dot; fringes rather paler. Beneath silvery-gray with slight smoky tinge, costa pale ochreous; primaries with prominent discal spot and t. p. line as on upper side, this latter shaded outwardly slightly with pale ochreous at costa followed by a smoky apical shade; secondaries with markings of upper side more distinctly repeated. Expanse 18 mm.

HABITAT. Silverton, Colo. (July 24-31). (McDunnough). 1 ♂. Type, Coll. Barnes.

Resembles in habits and general appearance the European members of this genus, which are *heliophile* and confined to the Alpine regions; it is closest to *rupestralis* Hbn., but differs in sufficient particulars, notably the markings of the underside, to warrant its separation.

ORENAIA TRIVIALIS sp. nov. (Pl. I, Fig. 14).

Palpi black, beneath white; head and thorax black slightly white sprinkled, latter with rear portion bordered with a white line; primaries smoky black with obscure maculation in black consisting of a basal dot, a rather straight t. a. line, a t. p. line, minutely dentate below costa and well incurved at fold, and black shades representing orbicular and reniform; an s. t. line is represented by a faint white shade line not reaching above vein 5; a black dotted terminal line; fringes smoky. Secondaries paler smoky with sprinkling of white scales outwardly, traces only of a dark subterminal line and dark dotted terminal line; fringes as on primaries. Beneath silvery gray with slight dot on costa marking inception of t. p. line and dark dotted terminal line. Expanse 19 mm.

HABITAT. Silverton, Colo. (Aug. 1-7). (McDunnough). 1 ♀. Type, Coll. Barnes.

Resembles in general marking *Titanio alticolalis* B. & McD., but differs in structure and in the pale underside.

MAROA Gen. nov. (Type *Maroa unicoloralis*).

Antennae of ♂ ciliate; labial palpi porrect, 3rd joint concealed, smoothly scaled; maxillary palpi filiform; front with corneous process consisting of a central conical tubercle and a short lateral truncate process in front of each eye; legs smoothly scaled; upper outside spur of posterior tibiae about half the length of inner spur, lower spurs short; vein 7 of primaries straight and well separated from 8 and 9.

M. UNICOLORALIS sp. nov. (Pl. I, Fig. 15).

Head, thorax, abdomen and wings pale creamy-white without maculation; beneath as above. Expanse 23 mm.

HABITAT. Phoenix, Ariz. (June 1-7); Arizona (June 1-7). 1 ♂, 4 ♀. Types, Coll. Barnes.

The species has possibly been confused with *succandidalis* Hlst. which has however an entirely different frontal structure and which seems confined to the Rocky Mountain region.

MICROCAUSTA BIPUNCTALIS sp. nov. (Pl. I, Fig. 16).

Primaries brown; t. a. line black-brown, angled outwards below costa, then slightly inwardly oblique and minutely waved to inner margin; *reniform consisting of two small superimposed orange-yellow* spots; t. p. line commencing at a point on costa above reniform, strongly outcurved to a point half way between reniform and outer margin, then parallel to same with slight incurve above vein 1; costa at apex of wing finely orange-yellow, bordered inwardly by a white line; basal half of fringes deep smoky, outer half white, cut at apex by a smoky streak; line of demarcation slightly undulate. Secondaries with outer margin excised above vein 4; deep smoky brown with paler fringes. Beneath pale smoky, secondaries with traces of a subterminal line; fringes as above. Expanse 10 mm.

HABITAT. Redington, Ariz.; Palmerlee, Ariz.; Santa Catalina Mts., Ariz.; S. Arizona (Poling). 9 ♂, 3 ♀. Types, Coll. Barnes.

The ♀'s are slightly deeper in color than the ♂'s especially on secondaries.

TITANIO LUTOSALIS sp. nov. (Pl. I, Fig. 17).

Primaries olive-green, suffused with white; a white basal dash above and along vein I, forked at extremity, inner margin white; white irregular shading descending from costa almost to junction of vein I and t. p. line; terminal area paler than median area, defining outwardly the course of the t. p. line which is shaded outwardly with white at costa and in lower half, rounded outwardly below costa to near outer margin it then is oblique to just above middle of inner margin where it forms a slight angle; white terminal line; fringes white, tinged with olive-green at base and cut by two dark lines outwardly. Second-aries dark smoky with darker terminal line; fringes smoky with pale basal line. Beneath dark smoky, secondaries rather paler. Expanse 20 mm.

HABITAT. Jemez Spgs., N. M.; Boulder, Colo. (July 14) (Oslar). 2 ♂, I ♀. Types, Coll. Barnes.

Resembles *helianthiales* Murt. but the t. p. line is not so white or distinct and recedes at once from the outer margin after the curve instead of running parallel to it for a short distance; secondaries are darker and without the white line.

TITANIO LAETALIS sp. nov. (Pl. II, Fig. 16).

Primaries light olive-brown shaded with white; an indistinct whitish ante-medial band; white shading along the veins beyond the cell to t. p. line; two slight dots in cell indicating orbicular and reniform, the former often lacking; t. p. line dark, angled slightly below costa, then gently curved and parallel to outer margin almost to vein I, where it becomes very oblique to middle of inner margin, bordered outwardly with white; slight dark apical shade; ter-minal dark line; fringes white cut by two dark lines. Secondaries white in ♂ with obsolescent dark subterminal line not reaching inner margin; dark broken terminal line and white fringes; in ♀ secondaries are smoky. Beneath primaries dark smoky with white fringes; secondaries white with smoky tinge along costa. Expanse 14 mm.

HABITAT. Redington, Ariz.; Babaquivera Mts., Ariz.; Denver, Colo. (July 7) (Oslar); Loma Linda, S. Bern. Co., Calif. (July, Sept.). 14 ♂, 2 ♀. Types, Coll. Barnes.

Belongs in same group as preceding species; the white secondaries should distinguish it readily.

TITANIO ALTICOLALIS sp. nov. (Pl. I, Fig. 18).

Head and thorax black; primaries deep smoky, slightly glossy, with macu-lation obscure; t. a. line black, slightly waved; orbicular and reniform repre-sented by small dark shades; t. p. line black, minutely dentate, exserted below costa, gently incurved in the fold, fringes smoky. Secondaries smoky crossed by an obscure blackish subterminal line; fringes smoky. Beneath smoky, paler than above with a blackish subterminal line common to both wings. Expanse 22 mm.

HABITAT. Silverton, Colo. (Aug. 1-7) (10,000 ft.) 1 ♀. Type, Coll. Barnes.

The species is close to *ephippialis* Zett. of which we have a pair from Dovrefeld, Norway, before us. It is deeper in color, especially the underside; the t. p. line is less squarely exserted and the fringes of secondaries are smoky, not whitish as in the European species.

POLINGIA. Gen. nov. (Type *Polingia quaestoralis*).

Antennae in ♂ very lengthily ciliate; labial palpi porrect, 3rd joint straight, long, pointed, clothed with hair; 2nd joint fringed beneath with long rough hair, maxillary palpi tufted at extremity with long hair; head with long tufts of hair at base of antennae; front with strong conical protuberance, sharply pointed; primaries with veins 3, 4, 5 well separated, 7 curved and approximate to 8 at base; mid-tibiae slightly haired.

P. QUAESTORALIS sp. nov. (Pl. I, Fig. 19).

Palpi, head, thorax and abdomen black, former intermixed with pale ochreous; primaries pale ochreous shaded costally and basally with black, the area to t. p. line sprinkled rather heavily with white scaling, giving a slight bluish tinge; t. a. line black, obscure, outwardly rounded; t. p. line hardly indicated except by a pale ochreous band that sharply defines the darker median area, parallel to outer margin from inner margin to vein 5, then rounded inwardly to costa; orbicular and reniform prominent, black, former oval, horizontal, latter lunate, vertical; a smoky patch on costa before apex continued faintly across wing parallel to outer margin; fringes smoky cut by a paler line; secondaries blackish crossed by a faint paler subterminal band broadening towards costa; outer margin narrowly pale ochreous; dark terminal line; fringes white. Beneath basal ⅔ of both wings to t. p. line blackish; outer portion pale ochreous; cell of primaries shaded with ochreous containing orbicular and reniform as on upper side; dark costal shade before apex; fringes as above. Expanse 19 mm.

HABITAT. So. Arizona (Poling). 1 ♂. Type, Coll. Barnes.

The genus is related to *Titanio*, differs however among other things in the strongly conical front and venation.

CHRISMANIA Gen. nov. (Type *Chrismania pictipennalis*).

Antennae of ♂ shortly ciliate; labial palpi porrect, 3rd joint concealed in long hair; 2nd joint roughly clothed beneath with long hair, maxillary palpi short, dilated slightly at extremity with scales; front with a long truncate prominence, the lower edge of which is concave in front with slight lateral points; mid-tibiae shortly haired; primaries with vein 7 straight and well removed from 8.

C. PICTIPENNALIS sp. nov. (Pl. I. Fig. 20).

Palpi olive brown above, whitish below; head, thorax and primaries deep olive-brown, latter with median area heavily scaled with pale whitish-ochre; t. a. line irregular dentate, slightly outwardly inclined, preceded by pale scaling, with rather prominent tooth below cubitus; t. p. line dark, dentate, slightly exserted opposite cell, strongly oblique inwardly to below reniform and then almost straight to inner margin; reniform a large prominent dark patch, almost resting on t. p. line; fringes smoky-brown. Secondaries pale orange-red with paler fringes slightly tinged with smoky. Beneath still paler orange-red than above, primaries with smoky fringes and apical shade; legs and body beneath whitish. Expanse 16 mm.

HABITAT. Redington, Ariz. (M. Chrisman) 1 ♂. Type, Coll. Barnes.

More closely allied to *Titanio* than the preceding genus, but easily recognized by its frontal structure.

PHLYCTAENIA RUSTICALIS sp. nov. (Pl. I, Fig. 21).

Head and thorax brown; primaries light gray, heavily sprinkled and shaded with liver-brown, especially along costa and in patches below apex and above anal angle, which are entirely this color; t. a. line indistinct, dark, outwardly oblique to below cell, then slightly waved to inner margin; orbicular a round brown blotch; reniform figure of 8, filled with brown with a white dot in lower portion; t. p. line brown, denticulate, arising from a small costal patch, evenly rounded opposite cell, forming a strong sinus above vein. 2, reaching to below reniform, bordered outwardly slightly with whitish; a terminal row of blackish dashes interrupted by pale ochreous dots; fringes checkered smoky and ochreous with smoky basal line. Secondaries smoky with fringes as on primaries. Beneath primaries pale smoky with outer portion of costa ochreous with 4 dark dots; orbicular, reniform and t. p. line as above; secondaries pale ochreous with discal dot and dotted postmedian line and very distinct terminal line composed of dark dashes; fringes pale ochreous. Expanse 22 mm.

HABITAT. Redington, Ariz. 3 ♂, 4 ♀. Types, Coll. Barnes.

Closest to *inquinatalis* Zell. from Labrador, which it much resembles in size, wing shape and maculation.

PYRAUSTA POTENTALIS sp. nov. (Pl. I, Fig. 22).

Palpi, head and thorax pale creamy; primaries whitish, sprinkled lightly with brown scaling; maculation fine, indistinct; t. a. line broken, perpendicular, indicated by dark scaling; orbicular a dark dot; reniform outlined in smoky, constricted centrally, slightly oblique; t. p. line fine, dark, almost perpendicular to inner margin from costa to vein one, then bent inwards and slightly upwards to below reniform and again straight to inner margin; terminal dotted dark line; fringes concolorous. Secondaries similar in color to primaries, rather whiter towards base with a dark subterminal line continuing t. p. line of primaries, forming a strong sinus below vein 2; faint discal dash which appears to be continued by the portion of the t. p. line following the sinus; terminal dark line; fringes whitish, cut by a dark line. Beneath essentially as above. Expanse 19 mm.

HABITAT. Redington, Ariz. 1 ♂, 1 ♀. Types, Coll. Barnes.

The front is rather roundedly prominent, but hardly sufficient we think to place the species in the genus *Metasia*. The pale color and the lack of prominent maculation are distinctive.

PYRAUSTA PSEUDORANALIS sp. nov. (Pl. I, Fig. 23).

Palpi brownish outwardly, white beneath; head, thorax and wings pale ochreous, the latter slightly sprinkled with brown scales; primaries with t. a. line dark, outwardly rounded, slightly waved; t. p. line dark, waved, straight from costa to vein 2, then bent inwards and upwards to below reniform and again straight to inner margin with slight angle on vein 1; orbicular round, secarcely indicated by faint dark outline; reniform lunate, outlined in smoky; terminal dark dotted line; fringes plumbeous with darker basal line. Secondaries with t. p. line dark, sinuous, with strong inward bend along vein 2, and slight outward bend on vein 5; terminal line and fringes as on primaries. Beneath paler than above with markings showing through. Expanse 24 mm.

HABITAT. White Mts., Ariz.; Palmerlee, Ariz.; Jemez Spgs., N. M. (July 8-15). 3 ♂, 5 ♀. Types, Coll. Barnes.

Remarkably like *Blepharomastix ranalis* Gn. in markings, but slightly larger, and differing in palpal structure, these being long and beak-like, considerably longer than in *ranalis;* the t. p. line is also slightly more waved in our new species; the color is much paler than that of the Arizona form of *ranalis*.

PYRAUSTA LUSCITIALIS sp. nov. (Pl. I, Fig. 24).

Palpi brown above, white beneath; head, thorax and abdomen brown, latter white-ringed; antennae in ♂ rather lengthily ciliate, in ♀ almost simple; primaries white, heavily scaled with deep brown leaving only traces of ground color in the shape of a few suffused patches in median area and beyond t. p. line; t. a. line lost in the dark suffusion; orbicular and reniform dark blotches, separated by a square white patch; t. p. line deep brown, dentate, strongly bent in below reniform almost to origin of vein 2, preceded and followed by small whitish patches, forming outwardly a broken row of white spots culminating in a larger patch above inner margin almost filling the bend of the t. p. line; fringes checkered brown and white at base, brown outwardly. Secondaries whitish, more or less suffused with smoky, with a broad deep brown marginal band, narrowing towards anal angle and scarcely attaining same; a postmedian slightly dentate dark line, bent inwards on vein 2 and separated from dark terminal area by a narrow whitish band; fringes as on primaries; beneath much as above, but paler. Expanse 25-28 mm.

HABITAT. Redington, Ariz. 1 ♂, 8 ♀. Types, Coll. Barnes.

Allied to *mustelinalis* Pack. in type of maculation, but paler, with the white patches much more prominent.

PYRAUSTA PILATEALIS sp. nov. (Pl. I, Fig. 25).

Palpi brown at sides, white beneath; head and thorax white tinged with brown at inception of primaries; primaries grayish white, very slightly sprinkled with brown scales, especially over median area; lines fine, brown, rather obscure, t. a. line, when visible, outwardly oblique with small inward tooth in fold; t. p. line rather closer to outer margin than usual, and semi-parallel to same, rounded inwardly at costa, slightly dentate in lower portion; reniform an obscure brown lunule; fringes tinged with brown. Secondaries very pale smoky, darker along outer border, with traces of a subterminal line most marked around vein 2; fringes white. Beneath whitish, immaculate. Expanse 20 mm.

HABITAT. Loma Linda, S. Bernd. Co., Calif. (April, June, July) (G. R. Pilate). 5 ♂, 3 ♀. Types, Coll. Barnes.

We take pleasure in naming this pretty little species after the collector, Mr. G. R. Pilate; the slight brown sprinkling on primaries is at times almost obsolete.

PYRAUSTA SARTORALIS sp. nov. (Pl. I, Fig. 26).

Palpi brown above, white beneath; head and collar brown, with a white line along the side of the eye; thorax brown, paling laterally; primaries deep creamy, suffused with brown, which at times occupies almost the entire median space; t. a. line indicated by a pure white oblique shade band, which separates the paler basal area from the browner median area; in well marked specimens the median area is bright brown with the exception of a small anterior costal portion, which is deep creamy and contains a brown dot indicating the orbicular, the cubital vein is also paler scaled, jutting into the dark area almost to t. p. line; this line is indicated by a pure white band, bent inwards at costa and then parallel to outer margin, making the median space twice as broad at costa as on inner margin; immediately before the outer margin and parallel to same is a fine white line shaded inwardly with brown, which at times occupies nearly the whole subterminal space but is usually confined to the costal portion; fringes creamy. Secondaries whitish tinged with brown outwardly and showing traces of an oblique line which descends from costa well before apex to outer margin just before anal angle. Beneath shiny white with markings of upper-side partly visible. Expanse 17 mm.

HABITAT. Loma Linda, S. Bernd. Co. Calif. (G. R. Pilate). 4 ♂, 5 ♀. Types, Coll. Barnes.

Belongs in the same group with the preceding species. In pale specimens the brown median area is reduced to a costo-apical patch and another below the cubital vein and the subterminal brown shading is almost wanting.

PYRAUSTA CORINTHALIS sp. nov. (Pl. I, Fig. 27).

Head and thorax olive brown; primaries with basal half olive brown, outer half rose-pink with an oblique white subterminal band parallel to outer margin; fringes pink. Secondaries smoky brown; fringes smoky, whitish outwardly. Expanse 18 mm.

HABITAT. Palmerlee, Ariz. (Biederman). 1 ♂, 2 ♀. Types, Coll. Barnes.

Allied to *volupialis* Grt. but without the inner white line.

NOCTUELIA TECTALIS sp. nov. (Pl. II, Fig. 10).

Head, thorax and abdomen light gray; primaries light gray, the median and terminal areas variably shaded with red-brown; basal portion of wing even light gray, this section bordered by a fine red-brown t. a. line, inwardly oblique, bent slightly inwards at costa, bordered by broad whitish shade; inner margin before t. a. line red-brown; median area largely red-brown with dark obscure lunate shade representing reniform; t. p. line fine, irregularly dentate, brown, from costa near apex slightly outwardly oblique to vein 4, bent inwards along this vein to below reniform, then slightly outwardly inclined to inner margin just before anal angle, bordered inwardly with white; terminal area shaded with red-brown with white terminal line; fringes long, gray-brown, basal half darker. Secondaries smoky brown with white spot on vein 2 near outer margin; fringes smoky on basal ⅓, outer portion whitish tipped with smoky. Beneath smoky, primaries darker than secondaries. Expanse 13 mm.

HABITAT. Babaquivera Mts., Ariz. (type ♂); Redington, Ariz.; Wilgus, Ariz.; Deming, N. M. (Sept. 1-7); Ft. Wingate, N. M. (May 24-30) (type ♀). 7 ♂, 4 ♀. Types, Coll. Barnes.

Belongs in the group with *puertalis* B. & McD. with roof-like frontal prominence.

NOCTUELIA ACHEMONALIS sp. nov. (Pl. II, Fig. 12).

Head and thorax olivaceous; primaries with basal third deep pink, extending narrowly along costa to near apex; remainder of wing pale olivaceous ochreous, bordered along outer margin narrowly with pink; fringes smoky. Secondaries deep smoky with concolorous fringes. Beneath pale smoky. Expanse 12 mm.

HABITAT. Redington, Ariz.; Deming, N. M. (Sept.). 7 ♂, 4 ♀. Types, Coll. Barnes.

The front of this species is roundedly protuberant, but not nearly so prominent as in typical members of the *Noctuelia* group; the species might possibly be referred to *Metasia* as the palpi have the 3rd joint rather hidden in the scaling. In general appearance the species resembles somewhat the *laticlavia* group of *Pyrausta*, but the palpi are not beak-like enough to refer it to this genus. In some specimens the basal pink area extends over half the wing leaving only a narrow band of ochreous.

NOCTUELIA PANDORALIS sp. nov. (Pl. II, Fig. 13).

Head, thorax and primaries pale olive-brown, the latter with basal area more or less suffused with pink, this area defined outwardly by slight white scaling not attaining costa; terminal area pink preceded by a white shade band; fringes pale ochreous. Secondaries dark smoky with slight whitish shade above anal angle and subterminally on vein 2; fringes whitish ochreous. Expanse 12 mm.

HABITAT. Deming, N. M. (Sept. 1-7). 2 ♀. Types, Coll. Barnes.

Close to the preceding, but more typically a *Noctuelia* in structure. Should easily be separated by the white scaling; in the cotype the inner margin of the terminal pink area is very irregular, showing a prominent incurve above fold; in the type this margin is almost straight.

HELIOTHELA COSTIPUNCTALIS sp. nov. (Pl. II, Fig. 14).

Palpi whitish; head black-brown ringed posteriorly with greenish white; collar dark-brown tipped with whitish; thorax and patagia black-brown, slightly sprinkled with whitish; abdomen similar in color with a white transverse ring on posterior portion of each segment; primaries deep black-brown sprinkled finely with whitish scales, with a small square whitish patch on costa ⅓ before apex; secondaries even black-brown; fringes on both wings dusky. Beneath as above. Expanse 16 mm.

HABITAT. Kerrville, Texas (Lacey). 1 ♂. Type Coll. Barnes.

Apparently best placed in this genus as the front is not protuberant, but evenly rounded. The species much resembles a small Noctuid of the genus *Spragueia*.

HELIOTHELA UNICOLORALIS sp. nov. (Pl. II, Fig. 15).

Palpi whitish; head, collar and thorax deep brown sprinkled with white; primaries deep black-brown with a slight bronze tinge, sprinkled with whitish scaling; an obscure oblique t. a. line is indicated by slight white scaling; on costa ⅓ from apex a few whitish scales show the commencement of a t. p. line; close to outer margin and parallel to same is a fine whitish s. t. line; fringes smoky with a darker basal line. Secondaries evenly dark black-brown; fringes paler. Beneath primaries smoky, tinged along costa with pale ochreous, with a slight ochreous dash in position of t. p. line; seocndaries smoky with basal half tinged with ochreous, this area defined rather sharply by a broad dark terminal band. Expanse 12-14 mm.

HABITAT. Babaquivera Mts., Ariz. (Aug.); S. Ariz. (Poling). 3 ♂'s. Types, Coll. Barnes.

Allied to the preceding, but smaller; differs also in the lack of a costal patch and the presence of a subterminal line.

CHLOROBAPTA Gen. nov. (Type *Chlorobapta rufistrigalis*).

Antennae in ♂ annulate and ciliate; *labial palpi minute,* porrect, scarcely extending beyond front; maxillary palpi short, slightly scale-tufted at extremity; *tongue obsolete;* front gently rounded, not prominent; primaries trigonate, apex rather pointed, veins 6 and 7 from below upper angle of cell, well separated, 2 and 9 stalked, 10 free; secondaries with *vein 5 from well above 4* and parallel to same, disco-cellular oblique, 6 from upper angle of cell, 7 and 8 stalked; spurs on mid and hind tibiae well developed and of equal length.

C. RUFISTRIGALIS sp. nov. (Pl. II, Fig. 17).

Primaries pale lemon yellow overlying a dark ground color, which shows through in basal 2/3 of wing and narrowly along outer margin giving a general greenish tinge to wing; an oblique red-brown subterminal dash from vein 2 to anal angle subparallel to outer margin; fringes white. Secondaries pure white, slightly shaded with smoky at apex in ♀. Beneath primaries deep smoky, whitish at apex and inner margin; secondaries white. Expanse 15 mm.

HABITAT. Pyramid Lake, Nevada (June 1-7). 1 ♂, 1 ♀. Types, Coll. Barnes.

The genus appears to be quite unique with its small palpi and obsolete tongue; this latter feature would in fact throw it into the *Schoenobiinae* according to Hampson's tables, but we do not care for this association as in other respects it is quite distinctly Pyraustid; probably it would be best placed after *Noctuelia.*

SCISSOLIA Gen. nov. (Type *Scissolia harlequinalis*).

Antennae slightly annulate and ciliate; front rather prominently rounded; labial palpi, short, porrect, evenly scaled, 3rd joint visible; maxillary palpi almost as long as the labial, tufted with scales at extremity; primaries with outer margin strongly incised below the apex above vein 6 and slightly a second time above vein 4; vein 7 straight and well separated from 8 and 9 which are stalked; 10 well separated. Secondaries with vein 5 *remote from 4 and parallel to same;* discocellular very oblique inwardly; 6 from below upper angle; 7 and 8 well stalked.

S. HARLEQUINALIS sp. nov. (Pl. I, Fig. 28).

Palpi, head and thorax ochreous, a black line at base of patagia and a blotch on rear portion of thorax; abdomen black; primaries ochreous shading into bright yellow outwardly; a black basal line; t. a. line black, broad, oblique to vein 1, then sharply angled and inwardly oblique to inner margin; t. p. line black, very oblique outwardly from costa to above vein 6, then perpendicular to above vein 4, running inwards from here almost to t. a. line and again outwards below vein 2 forming a long finger-like projection, then inwardly oblique and parallel to t. p. line above inner margin; the t. p. line is bordered outwardly by a narrow ochreous line which in turn is followed by a black shade line, rather diffuse, parallel to t. p. line and tending to fill up the finger-like projection, broadening slightly at costa and inner margin; preceding t. p. line is a costal dark blotch joined to a darker oblique dash representing the reniform, the space between this latter and the t. p. line filled with a blackish suffusion; an oblique white dash below apex of wing; black lines joining the t. p. line to the two incisions in outer margin of wing; below the 2nd incision three prominent black dots terminally; fringes at apex of wing white, cut by a black line and tipped with smoky, below first incision plumbeous with second incision slightly marked with white. Secondaries deep smoky with paler fringes bordered basally by a dark line. Beneath smoky with the exception of the apical and terminal portion of primaries which are yellow and show the maculation of the upper side reproduced. Expanse 13 mm.

HABITAT. Palmerlee, Ariz. 5 specimens. Type, Coll. Barnes.

The wing shape and markings remind one considerably of the Crambid genus *Prionapteryx*, but the structure is distinctly Pyraustid; the genus would probably fall in the neighborhood of *Mimasarta* Rag. and *Simaethistis* Hamp. this latter having vein 5 of secondaries in the same peculiar position as in our genus, which position according to Hampson is a very primitive characteristic.

PLATE I

1. *Glaphyria dualis* B. & McD.
2. *Ercta desmialis* B. & McD.
3. *Hedylepta futilalis* B. & McD.
4. *Sylepta brumalis* B. & McD.
5. *Blepharomastix santatalis* B. & McD.
6. *Hellula aqualis* B. & McD.
7. *Loxostege typhonalis* B. & McD.
8. *Loxostege roseiterminalis* B. & McD.
9. *Diasemia leucosalis* B. & McD.
10. *Diasemia zephyralis* B. & McD.
11. *Diasemia fenestralis* B. & McD.
12. *Gyros atripennalis* B. & McD.
13. *Orenaia coloradalis* B. & McD.
14. *Orenaia trivialis* B. & McD.
15. *Maroa unicoloralis* B. & McD.
16. *Microcausta bipunctalis* B. & McD.
17. *Titanio lutosalis* B. & McD.
18. *Titanio alticolalis* B. & McD.
19. *Polingia quaestoralis* B. & McD.
20. *Chrismania pictipennalis* B. & McD.
21. *Phlyctaenia rusticalis* B. & McD.
22. *Pyrausta potentalis* B. & McD.
23. *Pyrausta pseudoranalis* B. & McD.
24. *Pyrausta luscitialis* B. & McD.
25. *Pyrausta pilatealis* B. & McD.
26. *Pyrausta sartoralis* B. & McD.
27. *Pyrausta corinthalis* B. & McD.
28. *Scissolia harlequinalis* B. & McD.

Plate I

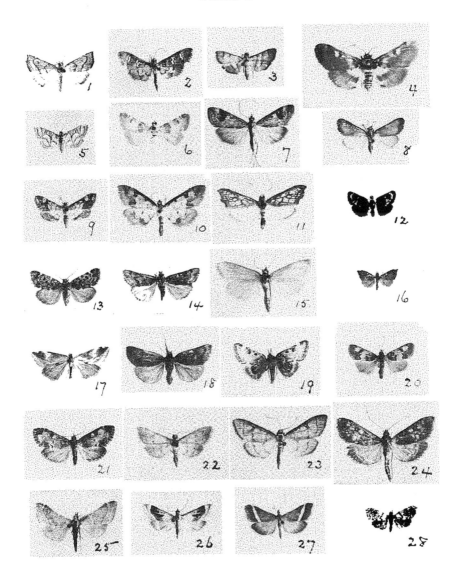

5050

PLATE II

1. *Evergestis funalis* Grt.
2. *Evergestis obliqualis* Grt.
3. *Evergestis eurekalis* B. & McD.
4. *Evergestis insulalis* B. & McD.
5. *Evergestis triangulalis* B. & McD.
6. *Evergestis obscuralis* B. & McD.
7. *Evergestis subterminalis* B. & McD.
8. *Evergestis vinctalis* B. & McD.
9. *Evergestis aridalis* B. & McD.
10. *Noctuelia tectalis* B. & McD.
11. *Evergestis lunulalis* B. & McD.
12. *Noctuelia achemonalis* B. & McD.
13. *Noctuelia pandoralis* B. & McD.
14. *Heliothela costipunctalis* B. & McD.
15. *Heliothela unicoloralis* B. & McD.
16. *Titanio laetalis* B. & McD.
17. *Chlorobapta rufistrigalis* B. & McD.

PLATE II

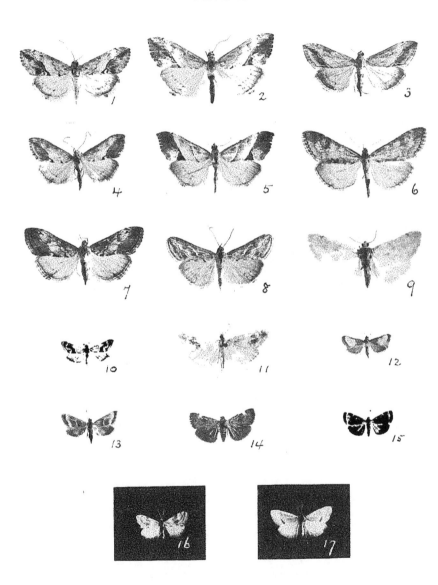

INDEX

PAGE

achemonalis B. & McD...... 243
allectalis Grt. 233
alticolalis B. & McD....... 238
angustalis Feld. 224
aplicalis Gn. 227
aqualis B. & McD.......... 228
aridalis B. & McD.......... 232
atripennalis B. & McD..... 235
bipunctalis B. & McD....... 237
brumalis B. & McD....... 227
catalaunalis Dup. 224
Chlorobapta B. & McD..... 245
Chrismania B. & McD...... 239
coloradalis B. & McD..... 236
consimilis Warr. 232
corinthalis B. & McD....... 243
costipunctalis B. & McD.... 244
desmialis B. & McD....... 225
dichordalis Hamp. 225
dualis B. & McD.......... 225
elegantalis Warr. 234
ephippialis Zett. 239
erubescens Hamp. 234
eurekalis B. & McD....... 231
extimalis Scop. 232
fenestralis B. & McD....... 235
funalis Grt. 228, 230, 231
futilalis B. & McD....... 226
harlequinalis B. & McD..... 246
helianthiales Murt. 238
indicata Fabr. 226
inquinatalis Zell. 240
insignatalis Gn. 224
insulalis B. & McD........ 229
laetalis B. & McD.......... 238
leucosalis B. & McD........ 234
lunulalis B. & McD......... 232
luscitialis B. & McD....... 241
lutosalis B. & McD........ 238
lygdamis Dru. 224
Maroa B. & McD......... 237
muiri Hy. Edw............. 236

PAGE

mustelinalis Pack. 242
napaealis Hlst. 228
oblectalis Hlst. 235
obliqualis Grt. 228, 230
obscuralis B. & McD........ 232
onythesalis Wlk. 224
pandoralis B. & McD....... 244
phoenicealis Gn. 224
pictipennalis B. & McD.....: 240
pilatealis B. & McD........ 242
plumbisignalis Fern. 235
Polingia B. & McD........ 239
potentalis B. & McD........ 240
pseudoranalis B. & McD.... 241
puertalis B. & McD........ 243
quaestoralis B. & McD...... 239
ranalis Gn. 227, 241
roseiterminalis B. & McD.... 233
rubralis Warr. 236
rufistrigalis B. & McD...... 245
rupestralis Hbn. 236
rusticalis B. & McD....... 240
santatalis B. & McD....... 226
sartoralis B. & McD........ 242
Scissolia B. & McD....... 245
simulatalis Grt. 228, 231
subterminalis B. & McD.... 230
succandidalis Hlst. 237
taeniolalis Gn. 224
tectalis B. & McD.......... 243
triangulalis B. & McD...... 229
trivialis B. & McD.......... 236
typhonalis B. & McD....... 229
undalis Fabr. 228
unicoloralis B. & McD.
    (Heliothela) 244
unicoloralis B. & McD.
    (Maroa) 237
vinctalis B. & McD........ 231
volupialis Grt. 243
zephyralis B. & McD....... 234
zephyralis Led. 224